Quantum Effects in Tribology

Quantum Effects in Tribology

Dmitry N. Lyubimov
Director for Science of Engineering Center
Lyubimov & Co. Ltd.
Shakhty, Rostov Region, Russia

Kirill N. Dolgopolov
Leading Engineer
Laboratory of Nanotechnology and New Materials
Rostov State Transport University
Rostov-on-Don
Rostov Region, Russia

With assistance from
L.S. Pinchuk

CRC Press is an imprint of the
Taylor & Francis Group, an **informa** business
A SCIENCE PUBLISHERS BOOK

CRC Press
Taylor & Francis Group
6000 Broken Sound Parkway NW, Suite 300
Boca Raton, FL 33487-2742

First issued in paperback 2021

ISBN-13: 978-0-367-78240-5 (pbk)
ISBN-13: 978-1-4987-6363-9 (hbk)

Visit the Taylor & Francis Web site at
http://www.taylorandfrancis.com

and the CRC Press Web site at
http://www.crcpress.com

Preface

"...we see unexpected things: we see things that are far from what we would guess – far from what we could have imagined. Our imagination is stretched to the utmost, not, as in fiction, to imagine things which are not really there, but just to comprehend those things which are there."

Richard Feynman, American physicist

Paradigm (Greek παράδειγμα "pattern, example, sample") *is defined as universally recognized scientific achievements that, for a time, provide model problems and solutions for a community of practitioners.*

Thomas Kuhn, *The Structure of Scientific Revolution*

This book is a continuation of the Tribophysics cycle (published by Springer as *Micromechanism of Frictiona and Wear. Introduction to Relativistic Tribology*) and Quantum Tribophysics. These books considered the changes faced by the atomic-molecular structure of a substance at friction and the associated micromechanisms of forming lubricating layers. They also tried to develop a mathematical formalism relevantly describing the stages of tribosystems evolution resulting from the ongoing tribotransformations.

Any theoretical model of modern physics should be based on quantum mechanics as a fundamental physical concept that describes the laws of the microworld. In this book, we introduce the quantum mechanical formalism into the presentation of the basics of tribology—from the revision of views on the nature of friction to the description of its special chapters reflecting the tribomutations, the fields generation as a result of frictional interaction of particles, 'birth' and 'destruction' of tribosystem elements at the micro-, meso- and macrolevels of its evolution.

In the methods of quantum theory, when adapted to the concepts of friction processes, we actively use the publications of the greatest physicists who created the modern picture of our being—from Einstein and Bohr to Penrose and Hawking.

In order not to become like the Swift's commentators[1] in interpreting the ideas of classics who built the modern physical worldview that formed the theoretical basis of this book, we apply the method of a broad quotation of the scientists whose works

[1] In his numerous travels, the main character of Jonathan Swift, Lemuel Gulliver, finds himself on the island of Glubbdubdrib ruled by a magician-necromancer who is able to communicate with deceased heroes. With his help, Gulliver meets, among others, Homer and Aristotle: "... I soon discovered that both of them were perfect strangers to the rest of the company, and had never seen or heard of them before; and I heard a whisper from a ghost who shall be nameless, that these commentators always kept in the most distant quarters from their principals, in the lower world, through a consciousness of shame and guilt, because they had so horribly misrepresented the meaning of those authors to posterity."

are included in the contents. Talking about quantum models, we try not to distort the meaning put by their creators.

We accept this form of information transfer and the layout of the book's chapters so that the presentation of the selected topic of quantum-mechanical triboprocesses nature is acceptable from both the classical positions of tribology and new ideas in physics. Since quantum theory is an extensive fundamental scientific discipline, it is impossible to include all its methods in one book. Hence, the necessary additional information from each chapter of the scientific disciplines embedded in the book are collected in the form of short definitions, supplements, digressions and explanations, which facilitate a compact presentation of such vast material.

The contents of the book center around the development of Bohr's theory on the nature of the measurement of physical parameters with respect to friction units and the impact of this process on the properties of the tribosystem. Very useful are the ideas about the non-locality of quantum objects, reduction of the wave function and the elements of secondary quantization formalism. The evolution of friction units is depicted in the form of spatial transitions of a quantum tribosystem: the birth of the elements of the tribosystem structure out of the physical vacuum in the friction process, the formation of friction subsystems by those elements; the interaction between the subsystems, through interpenetration of non-local connections, leading to the formation and transformation of the tribosystem as a whole. Using the chosen approach, one may obtain not only the quantum microparameters that reveal the nature of the fundamental processes that base the frictional interaction, but also derive the relations to evaluate the quantities traditionally accepted in tribology. Use of the formalism of quantum theory allows reliable and logical bringing together of mechanical and atomic-molecular friction models with the ideas of modern tribology, such as tribonic model of frictional interaction of solids, which is given in detail for the first time in tribological literature.

The book shows that the application of the laws of quantum mechanics allows creation of an integrated friction model or even a unified tribological theory in general scientific sense. We hope that the first such essay in tribology will find its supporters and yield in both solving theoretical problems and popularization of tribology and related scientific areas among the physics audience.

With profound respect to all who open this book.

Contents

Chapter 1

Problems in Creation of the Generalized Physical Model of Friction and Wear*

The centuries-old history of the study of friction and wear was marked in the second half of XX century by the formation of a specialized discipline – tribology [1]. Unfortunately, that did not lead to the development of a unified doctrine, which would make it possible to describe the processes of friction, lubrication and wear. This is evidenced, first of all, by the definitions of tribology as a science.

1.1 Subject and Methods of Tribology

For the first time the term 'tribology' was proposed by a group of British experts in a report to the Parliament on lubrication problems. Its contents ran as follows: "*Tribology is a science and technology of interacting surfaces in relative motion, as well as related phenomena and their practical consequences*" [2].

*Work carried out with the financial support (Project No. 14-29-00116) the grant of the Russian Science Foundation, K.N. Dolgopolov (Rostov State Transport University) participated in the project.

According to this definition, tribology consists of two parts: the first part is actually a science concerned with the study of physical and chemical phenomena that occur upon contact of friction surfaces; the second part is the technology of optimization of frictional interaction in tribosystems and its application in practice.

Russian sources often use the basic definition given in [3], which could be translated into English as follows: "*Tribology (tribonics) is a science concerned with a study of friction, wear, lubrication and interacting contact surfaces in their mutual displacement.*"

Thus, tribology is, first of all, a study of the processes of friction, wear and lubrication in friction units characterized by physical and chemical mechanisms that 'work' on atomic and molecular, nano- and micro-levels (adhesion, tribochemistry, triboplasma, etc.).

The duality of this term is clearly visible in other definitions, for example:

"*Tribology is a science of friction and the processes that accompany friction*" [4].

"*Tribology is a science and technology of bodies' interaction processes in their relative motion*" [5]. This definition, like the first one, names the two groups of problems solved by different methods of tribology. It looks a little odd to oppose the science and technology which, therefore, is not considered a science. The ending of the formulation is imperfect because it hints at the similarity of the subjects of research in tribology and, for example, celestial mechanics, which also studies interacting bodies in their relative motion. However, the interactions of celestial bodies are studied based on relativistic ideas.

In the fundamental tribology articles collection [6] that contains articles by specialists from the CIS states and the USA, Peter J. Blau led the definition commonly used in the United States: *Tribology is Friction, Lubrication and Wear.* It does not have 'transition steps' from the clear-cut subject of tribology to the required unknown, i.e. tribological parameters of friction units. These steps may be determined by analyzing the areas of tribology listed in the article by V.A. Bely, N.K. Myshkin from the same articles collection: tribology is triboanalysis (i.e. tribomechanics, tribophysics and tribochemistry) + tribonic material science (structure and properties of friction materials and their changes during friction) + tribotechnics (parameters of tribological conjunctions) + tribotechnology (tribological conjunctions performance control) + tribonic computer science (processing and storing the information about tribosystems).

Such a specification makes clear the division of tribology into the physical and chemical part that studies micro-processes in a frictional contact (triboanalysis, tribonic material science and tribotechnology) and the technical part, the area of interest of which is limited to mechanical interactions of friction parts and macro-characteristics of friction units (tribotechnics, tribomonitoring, tribonic computer science). Of course, this is only a rough division that does not take into account the plurality of tasks in each direction, as well as the current trends in technical sciences to conduct research on both micro- and macro levels. The latter, however, only shows the relationship of the tribology areas that have adjacent areas of study, but says nothing about the unity of the methodological approach to the study of mechanical interactions in a tribosystem and physical and chemical processes and phenomena in a frictional contact.

Let us analyze the reasons for this situation in tribology and the ways out of it.

1.2 Classical Structure of Friction Science

The duality of tribology is a consequence of the evolution of ideas about friction as a natural phenomenon. The binomial friction law discovered in 1778 by a French physicist and engineer Charles-Augustin de Coulomb represented the force of friction as a sum of cohesions A of the bodies that do not depend on the external load and the product of the friction coefficient f^* (Coulomb considered it permanent) and the value of the normal load in the friction pair: $F_f = A + f^*N$.

According to the molecular-kinetic theory of friction by I.V. Kragelsky (1939), along with the deformation component of the friction force F_D generated by deformation of microroughness in contact displacement of the frictional surfaces, the adhesive or molecular component F_A that arises as a result of adhesion of the frictional surfaces makes its contribution on the force of friction:

$$F_f = F_D + F_A \tag{1.1}$$

Although the friction model by I. Kragelsky is considered fundamental, there are still attempts to represent the friction force as a result of interaction of atoms and molecules in friction materials. One example is a formula that represents a sum of the adhesive F_a and the cohesive F_k components of this interaction [7]: $F_f = F_a + F_k$.

This means that the adhesive interaction occurs between the friction surfaces of the parts in dynamic contact, while the cohesive one occurs between the material's particles in the surface layer of each of these parts.

Base Friction Models: Of course, the development of friction science is not limited to changes in notation of the binomial friction law. Tribology is a young and rapidly developing discipline, the results of which are in very high demand by industry all over the world. However, either because of predominance of applied orientation of tribology and the public pressure or because of complexity of friction as a phenomenon and a subject of study, there is still a lack of theory in the friction science that would make it possible to describe both mechanical interactions of friction parts and the physical and chemical phenomena and processes that occur in dynamic contact using a unified methodology.

This situation is not due to lack of ideas and original friction models; they are exactly developed in quite a large amount. The piquancy of the situation prevailing in tribology is the existence of well-founded but poorly related, sometimes even conflicting opinions on the nature of friction, wear and lubrication, which in some cases determine the antagonistic relationship between the different scientific schools. The main contradiction is a conflict of views on friction as a mechanical process and the concept that considers friction as a result of atomic-molecular interaction that occurs between the friction surfaces and the adjacent surface layers of friction parts. The extreme expressions of these points of view, according to A.S. Akhmatov, the reputed Russian tribophysicist [8], are the mechanistic friction model by G.I. Yepifanov, a Soviet tribologist and the model that was simultaneously and independently proposed by A. Cameron (American physicist and tribologist) and B.V. Deryagin (corresponding member of the Academy of Sciences of the USSR, specialist in surface phenomena).

The views of G.I. Yepifanov developed on the basis of the results of carefully prepared experiments carried out in his laboratory to establish the relationship between friction and adhesion forces in frictional contact. In the first series of experiments, the metal samples with smooth, well-cleaned contact surfaces were used; their contact form strong adhesive bonds. In the second series, the samples were made of materials 'that do not adhere to the fresh surface of the metal counterbody' (ivory, coconut shell, etc.). According to the results of the experiments, a binomial equation for the friction force was derived [9]:

$$F = \theta_0 S + kN \tag{1.2}$$

where θ_0 is the shear strength of friction material at the normal stress; $\sigma = 0$; S is the field of contact of the samples; $k = \delta\theta/\delta\sigma$ is the shear strength rate at the change of normal stresses; N is the normal pressure.

G. Yepifanov concluded that friction forces are entirely determined by the shear strength of friction materials and does not depend on the adhesion in a movable contact.

The boundary friction theory by A. Cameron suggests that during sliding, the tribosystem periodically passes through the maximum (U_1) and minimum (U_2) positions of the potential energy. The friction force work F is

$$F_x = U_1 - U_2 \tag{1.3}$$

where x is the distance between maximum and minimum energy positions.

Cameron introduced a simplification $U_2 = 0$ and derived an equation for friction force [10]:

$$F = U_1/x \tag{1.4}$$

The basic assumption of the model ($U_2 = 0$) led to objections by B. Deryagin and A. Akhmatov. As a result of the discussion that arose after Cameron's report in the Mechanical Engineering Institute of the Academy of Sciences of the USSR in 1960, Cameron was forced to make a significant refinement in the first version of his model. However, he was the first to construct a theory of friction on the basis of the precise laws of atomic interactions, which played a significant role in theoretical tribology.

The friction model by B.V. Deryagin implies that the dynamic contact of solid materials passes through three stages: a) elementary act of friction, b) single crystals friction and c) polycrystalline bodies friction [11].

a) When perfect single crystals with the same orientation of crystal lattices are in contact, the surface atoms of the lattices interact. An elementary act of friction is an interaction of electron shells of the atoms that belong to different lattices. The reason for such a representation is as follows. The atomic structure of the crystal planes determines the discontinuity of their force fields. The configuration of the latter in single crystals contact zone is adopted as a system of contacting spherical surfaces of equal radius. The slipping under these conditions is due to energy consumption and an emergence of resistance forces, i.e. the friction forces. The author neglects thermal vibrations in the atoms, which is equivalent to the exclusion from consideration the temperature function of friction.

The displacement of each atom of one crystal lattice with respect to the other is followed, like in the Cameron's theory, by surmounting a number of energy barriers of the conjugate lattice force field. In this process, an elementary act is a displacement of the point on spherical surface $z = \phi(x, y)$ with its transition from the initial equilibrium position to the limit non-equilibrium position followed by a jump to a new equilibrium position. The static friction coefficient is equal to the maximum value of the derivative dz/dx in the limit unstable position of the selected point.

b) When regard to the transition from the elementary act of friction to single crystals slip, the author neglects the tangential components of attractive forces. The latter has a large work range and decreases with distance more slowly than the repulsive force. This leads to a low height of their potential barriers. These assumptions made it possible to turn to consideration of the slip of polycrystalline bodies simple.

c) The friction of polycrystalline bodies is a statistical result of the interaction of a large number of pairs of monocrystalline areas. They have random orientation in space and different crystal structures. B. Deryagin divides friction pairs into groups with the same structure and orientation of the areas. Within the group, the disturbance of crystal orientation in time is taken into consideration by a 'phase difference'.

Despite the disadvantages of these assumptions, the model by B. Deryagin is still the standard of rigorous physical approach to the description of friction in theoretical tribology.

Hardy Theory: An important step in the development of tribology was developing ideas about the presence of a third phase between the friction surfaces. Friction of solids mostly occurs when there are adsorption films on the friction surfaces. English physicist and chemist W.B. Hardy introduced the terms 'boundary lubrication' and 'boundary state' into the tribology system of conceptions [12]. Friction at boundary lubrication covers a broad and technically important area of physical and chemical conditions of frictional interaction, thus occupying an intermediate position between 'dry' and fluid friction. Hardy performed the first systematic study of boundary friction that retains its value to this day. It should be recalled that Hardy's student, F. P. Bowden, the founder of the Tribological School in Cambridge, and his colleague D. Tabor, developing Hardy's ideas, created the adhesive friction model [13].

Hardy studied static friction of glass, steel and bismuth in homo- and heterogeneous pairs as a function of molecular weight of lubricants – paraffins, fatty acids and alcohols. As indenters, he used spherical lenses that allowed observation of the lubricating layers and argued that the friction factor f was a linear decreasing function of molecular weight M of the lubricant:

$$f = a - bM \tag{1.5}$$

where a and b are empirical coefficients.

It appeared that the relation 1.5 is valid for all combinations of materials used in the experiments. In the absence of lubricant $(M = 0), f = a$, i.e. the static friction coefficients depend only on the properties of the solids. If a friction pair of identical materials is characterized by a constant a_1 and a pair of other materials is a_a, then for the pair of the first and second materials, $a = (a_1 + a_2)/2$. For the friction pairs lubricated with substances of different homologous series,

$$f = a_0 - d - bM \tag{1.6}$$

where a_0 depends on the nature of the solids; d and b only on the chemical structure of the lubricants; $bM = c(N - 2)$; c is an empirical constant; N is the number of atoms in carbon.

Hardy writes: "Apparently, at sufficient chain length, the field disappears, hence the friction is absent. It really is what we observed in the following form: with the least tractive force we could realize, there was slippage." He noted that the measurement of static friction in values close to zero is difficult, since the transition to the slip occurs abruptly and its registration accuracy is not sufficient.

Dependencies 1.5 and 1.6 enable us to determine the molecular weights of the lubricants that ensure minimal friction. The best value $M = 180..190$ in the class of fatty acids, 320 for alcohols, 400 for paraffins. Further experiments confirmed Hardy's calculations.

(a) G.I. Yepifanov

(b) B.V. Deryagin

(c) W. Hardy

Situation Assessment: The authors did not strive to give a complete retrospective of ideas, methods and models of tribology in the preceding paragraphs. This topic is exhaustively reflected in [5, 6] and in a fundamental monograph by D. Dowson [14]. The above are just significant signs of a few friction theories and models that are in use up to date and important for further disclosure. The extreme positions of their authors absolutely obviously illustrate the remarkable situation in tribology, which sets it apart from other natural and applied sciences.

In friction theory, there are dozens of well-developed models existing at the same time, based on finely set and carefully executed experiments. However, their results make it possible for highly qualified professionals to treat the same natural phenomenon – friction – in different ways. The authors of various models rarely compromise, thus remaining self-sufficient within their own views.

The fundamental attempt to combine the basic mechanisms of friction in one model prompted I.V. Kragelsky to create a fundamental molecular-mechanical theory of friction [4, 5, 6, 15]. It is based on binomial law 1.1, according to which the frictional force has deformative and adhesive components, the latter being proportional to the actual contact area. Otherwise, i.e. on equation forms and the parameters containing in them, this is a mechanical model. The apparatus of Newtonian mechanics that formed the basis of molecular-mechanical theory, even being developed by modern methods of mathematical physics, cannot be considered a theoretical core around which one may build a unified theory of tribology.

The history of modern tribology knows several attempts at forming such a core.

B.I. Kostecki, the founder of Ukrainian school of tribology, developed a thermodynamic concept of friction [16]. Currently, it has received an additional impetus in development in connection with active application of non-equilibrium processes thermodynamics in tribology [17]. The equations of energy balance and thermodynamic functions describing the frictional processes are parameters of macroscopic (i.e. consisting of a large number of particles) physical systems. Micromechanisms of thermal phenomena are explained in molecular physics, using macro-dynamics representations. This makes it possible to represent the frictional interaction as a result of the emergence and destruction of adhesive bonds between the friction surfaces. However, the global generality of approaches to the disclosure of the essence of fundamentally different natural phenomena peculiar to thermodynamics is not always useful in friction theory. It makes it difficult to quantify the most important for tribology mass transfer parameters in a tribounit, appearance and interaction of physical and force fields, kinetics of tribochemical reactions, etc.

A more differentiated approach to the choice of the theoretical core of tribology may be based on the many (if not all) of the processes and phenomena that accompany the friction and are classified as macroscopic quantum effects. Features of quantum mechanics are peculiar to this set of phenomena, which are manifested in behavior of macroscopic objects [18]. In other words, for a description of several macroscopic systems, one may apply the equations that explicitly include the characteristic value of quantum mechanics, the Planck constant h. This is typical for such phenomena, as superfluidity of liquid helium, superconductivity, Josephson effect, quantum Hall effect.

In quantum mechanics, the state and motion of a system are described by a wave function that has no classical counterpart. In macroscopic quantum processes the wave function is shown as a registered macroscopic parameter of the system. This is possible if the system has a large number of particles in the same quantum state. In its simplest form, this situation is realized in the classical electromagnetic field, which is in fact a combination of a large number of photons in close quantum states [18]. Similar phenomena also occur in cases when in one state there are a large number of 'ordinary' particles, e.g. atoms and ions, the number of which, in contrast to the number of photons, is constant.

One of such phenomena is triboplasma, a collection of charge carriers, which are released by friction from high-temperature areas of microroughnesses located in movable contact and interact via the Coulomb forces. In contrast to the gas plasma, all the components of which (electrons, ions, atoms) are mobile; ions and atoms of friction material make only small oscillations about their equilibrium positions. Only a fraction of the charges serve as mobile charge carriers that form the triboplasma. They move in a self-consistent field under conditions of, firstly, strong interaction with the field of friction material crystal lattices that form their energy spectrum and secondly, collisions with impurities and defects in the surface layer of the lattices. These collisions serve as a strong relaxation channel of triboplasma generation missing in a gas plasma [19].

The phenomenological model created by German physical chemists, P.A. Thiessen, K. Meyer and G. Heinicke, presents the formation of triboplasma as follows [20]. A pair of microasperities belonging to different friction surfaces experience elastic or plastic deformations in contact. In a very short period of time of a single contact of microroughness pair ($10^{-7}..10^{-8}$ s), a large amount of energy is released in the contact zone. Heating microzones are formed, the area of which does not exceed $10^{-4}..10^{-2}$ of the nominal contact area. The heat energy emitted may so exceed the specific melting heat of friction materials that it would lead not only to their melting, but also to sublimation. Then the friction materials evaporate at the contact points and move into the triboplasma state as ions and electrons. Japanese physician, K. Nakayama, in 2002, experimentally confirmed the generation of triboplasma at rotational friction of a diamond sphere on a sapphire disk [21].

Along with triboplasma, the category of macroscopic quantum effects should include triboelectrization, frictional heating, friction noise, the phenomena of selective transfer and abnormally low friction in a vacuum which were recognized as discoveries in the USSR [22] and other processes that contribute to the self-organization of tribosystems. Starting with a high level of triboexcitation, the collective motion of the third body particles leads to a transition of the tribosystem into the state of dissipative space-time structure, which may be described in a probabilistic form, using the mathematical apparatus of quantum mechanics.

Thus, the modern science of friction supposes the simultaneous existence of related but independent and even contradictory theories and models of friction and wear. The problems of their unification into a single unit are described in the next section.

1.3 Paradigms of Tribology

At each stage of development of any scientific field, a system of representations on the subject of the study is created, which at the given moment of history seems to be so obvious as to be identified with so-called common sense. The latter becomes the ideological basis of the paradigm; a scientific theory that constitutes the fundamental basis of the scientific discipline.

Paradigm and Science Development: Paradigm is a scientific theory embodied in a system of concepts that express the essential features of reality. It brings together the achievements in a given field of science recognized by society and elaborates the way to organize the scientific knowledge, which for some period provides the scientific community and all of humanity the ideas about the picture of the world in general and the nature of specific phenomena, in particular. The paradigm helps researchers to formulate the problem and decide how to solve it. According to the Ukrainian theoretical physicist, O.O. Feygin: "The paradigm provides us with something like a grid to facilitate the understanding of the deep meaning of the events. With its help, out of the many possible explanations, we select the most probable ones, rejecting the remaining as unscientific. Science is very economical in its explanations and always tries to use a minimum of assumptions" [23].

The paradigm methodology makes it possible to overcome the difficulties that inevitably arise in scientific research and record changes in the structure of knowledge that occur due to accumulation of experimental results. The paradigm defines the 'sustainability' of scientific theory to the shocks related to the discovery of new phenomena that to some extent change the previous ideas. In such cases, it becomes obvious that the private paradigms of individual scientific fields are closely related to the philosophical concept that reflects the history of human interaction with Nature.

The paradigm is peculiar to each scientific discipline. Thus, the paradigm of classical electrodynamics consists of the following achievements: a) the phenomenon of electromagnetic induction discovered as a result of Faraday's experiments; b) vector analysis as the basis of the mathematical tools of electrodynamics; c) the Maxwell's equations written with the help of it. Other examples of paradigm: the Einstein field equations in the general theory of relativity, the Lagrange and Hamilton equations in classical mechanics, etc. These equations formally display a limited number of common features of a scientific field: the equivalence principle in the theory of relativity; the principles of complementarity and correspondence in quantum mechanics; the principles of virtual displacements and least action in classical mechanics.

The works by Leonardo da Vinci, dating from the second half of the 15th century, are recognized as the basis of modern tribological paradigm. He found that solid bodies with a rough contact surface have a higher resistance than those with a smooth surface and discovered a law, according to which the friction force is proportional to the load, i.e. $F_f = fN$. The ideas of Leonardo correspond to a remarkable period of human development – the Renaissance, which gave birth to revolutionary ideas of the heliocentric world system by Copernicus, the foundations of modern mechanics by Galileo, the philosophy of Descartes based on the dualism of body and soul.

These achievements anticipated the appearance of Newton's genius and led to the formation of the mechanism, the one-sided ideological principle that explains the development of Nature and society by the laws of a mechanical form of the matter motion. According to the metaphysical view, natural phenomena were seen as a result of a coupling of elementary cog-wheels, threads tension, springs elasticity, etc., and the world was seen as a kind of 'clockwork', where all processes were made in a similar manner. The views of Leonardo on friction were absolutely consistent with the general mechanistic paradigm of the late Middle Ages and the beginning of the modern era.

However, the stabilizing effect of the paradigm on the development of scientific concepts and function of gradual accumulation of empirical facts are not always a positive factor, but at a certain time become its Achilles' heel. The point is that the wonderful and diverse world from time to time throws new facts to the researcher, giving rise to formal logical contradictions in the usual paradigm called paradoxes.

Paradoxes in Science: In logic, a paradox is a contradiction resulting from a reasoning, logically correct at a first glance, leading to mutually contradictory conclusions. A paradox arises when two mutually exclusive propositions are equally provable. In scientific work it means, firstly, that in the accepted paradigm there is no criterion to establish the truth with respect to the controversial factor and secondly, it is the first signal of the coming change of paradigm. This conclusion is consistent with the opinion of well-known American physicist and Nobel Prize winner, R. Feynman: "A paradox is a situation which gives one answer when analyzed one way, and a different answer when analyzed another way, so that we are left in somewhat of a quandary as to actually what should happen. Of course, in physics there are never any real paradoxes because there is only one correct answer; at least we believe that Nature will act in only one way... So in physics, a paradox is only a confusion in our own understanding" [24].

A bright historical example of a paradox are Zeno's paradoxes. Zeno was a Greek philosopher (490-430 BC), whom Aristotle considered the founder of dialectics as the art of understanding the truth by dispute. Zeno is known for his famous paradoxes that justify the impossibility of motion, multiplicity of things, etc. Here is an example of Zeno's arguments. "If everything when it occupies an equal space is at rest, and if that which is in locomotion is always occupying such a space at any moment, the flying arrow is therefore motionless" [25]. The riddle of the Zeno's paradoxes that lasted for millennia was resolved with the advent of mathematical analysis created by Isaac Newton and G. Leibniz. The presence of such arguments about the nature of motion ultimately determined the form of the laws of classical mechanics.

The confusion, Feynman says, indicates the beginning of the destruction of the old paradigm, the tools of which are no longer able to adequately describe the reality that appears beyond the ideas about it. Einstein called this situation an 'act of surprise': "Act of surprise apparently arises when the perception comes into sharp conflict with the established concepts in our world" [23]. This world of concepts determined by the old classic paradigm reflects the limitations and the archaism of its positions, which enable establishment of the readiness of the scientific community to the paradigm shift.

Requirements for a New Paradigm of Tribology: Classical mechanics, which had been the basis of the modern paradigm of tribology, has been subjected to numerous revisions over the past 300 years. New concepts appeared: integrals of motion, energy, momentum, action, laws of conservation, equations derived using advanced mathematical techniques. Modern classical mechanics bears little resemblance to its original version stated by Newton in his Principles of Natural Philosophy. Then the new physical concepts were created, according to which classical physics was only a modest special case of quantum relativistic physics. Unfortunately, tribology, especially in its theoretical part, responded poorly to these changes.

Of course, since the views expressed by Leonardo, much has changed in the theory of friction. The appearance of the binomial law and its evolution from Coulomb to G.A. Tomlinson (English tribologist and the first developer of molecular adhesion friction theory) and Deryagin indicate that a concept prevails in modern friction theory, according to which "... all triboeffects appear in the macro scale, but the phenomena that cause and accompany them occur on micro levels" [7]. This assertion is based on a large experimental basis. On the other hand, the positions of the opponents of the atomic-molecular friction theory are equally strong [4, 5, 6].

The main reason appears to be as follows. Almost all the publications devoted to friction micromechanisms are purely experimental. Their theoretical interpretation is often limited to qualitative assessment of the curves progress, using concepts of physics and chemistry. Ethical considerations do not allow us to give examples of such constructs that may not be considered serious theoretical models of atomic-molecular friction.

Only the energy models look decent in this background, as evidenced by the apparent success of this direction of tribology [3, 4, 17].

A major achievement of the modern wear theory became the model proposed by N.M. Alexeev and M.N. Dobychin [6]. The theory presents a wear process as a result of natural sequential change of mechanisms (and therefore, the physical models) of destruction of the surface layers of friction parts. The successive partial models describe the fractures that occur at different scale (structural) levels – from disclocations up to visibly obvious fracture fragments. Unfortunately, this theory does not consider the frictional interactions at the atomic-molecular level.

So the attempts to combine all the models proven by experiments in a single rigorous system failed even in terms of the fundamental molecular-mechanical friction theory. This means that there is a paradox in modern tribology and requires new ideas for at least restoring the adopted paradigm.

A paradigm shift occurs, as a rule, during scientific-technical revolutions: Leonardo da Vinci and Coulomb – the first industrial revolution, Bowden and Tabor – the second industrial (electrical) revolution, Kragelsky – model (nuclear) revolution (the names are chosen at random, other preferences allowed). Naturally, new theories meet eligibility criteria, according to which the elements of previous theories based on proven facts, being changed, are included in new models as their less common (private) components. That is, the new paradigm does not reject its secular heritage, but absorbs it after being creatively rethought, thus presenting as an integral part. The eligibility makes it possible to build a new theory not from scratch but on the basis of the previously established knowledge platform using the fundamental interdisciplinary principles of scientific logic. With regard to the current state of tribology, the most relevant principles are 'overcoming the taboo' and 'coincidence of the evolutionary path'.

The principle of 'overcoming the taboo', i.e. violations of the prohibitions to perform certain actions, has long been used in physics and in the formulation of the Russian scientist, V.S. Barashenkov, reads: "...everything that is not contrary to the known laws of Nature has a right to exist" [26].

The principle of 'coincidence of the evolutionary path', also known as the analogy principle, is observed when developing any new model of natural processes – physical, chemical, biological or social [27]. Let us explain its essence through the following example.

D.N. Garkunov (together with A.A. Polyakov) paid attention to the resemblance in the structural changes observed in the friction units during the selective transfer mode with the evolution of biological species. He reached this conclusion despite the fact that "for a long time there existed something like a contradiction in the laws of development of animate and inanimate nature. The main law of thermodynamics by Clausius predicts the growth of entropy, i.e. disorder in a closed system. The evolutionary Darwin's theory, on the contrary, showed that the selection is based on the increase of the degree of organization of biological systems" [17]. This analogy was discovered by D.N. Garkunov before the Soviet scientific community gained access to the results of the Nobel winner in the field of thermodynamics of irreversible processes, the Belgian physicist, I. Prigogine. The theorem mathematically proved by Prigogine states that when the external conditions prevent the achievement of the equilibrium state by the system, the steady state of the system corresponds to the minimum entropy production. The similarity in evolutionary paths of natural selection and the wearlessness effect is a consequence of Prigogine's theorem.

One may conclude that the requirements for the future paradigm of tribology are contradictory. It must preserve the advantages of the molecular-kinetic theory and must not be in conflict with the reliable data used as an argument in a number of mutually exclusive friction models. There is a danger that this symbiosis will lead to an inconvenience in the application monster that is 'loaded with options' from various mechanisms, models and formulae but has no unified theoretical core.

In this situation, it is advisable to take into consideration the opinion of the English physicist, S. Hawking (one of the developers of the theory of black holes): "A model is a good model if it: 1) is elegant; 2) contains few arbitrary or adjustable elements; 3) agrees with and explains all existing observations; 4) makes detailed predictions about future observations that can disprove or falsify the model if they are not borne out" [27]. Hawking emphasizes that there is no reality that does not depend on the general picture of the world and cannot be described as a theory. Therefore, science offers to the public the so-called model-dependent realism: any theory is a model (often mathematical) of a natural phenomenon as well as a set of rules, that connect the elements of the model with the observations.

Stephen Hawking

1.4 The Analogy Principle as a Step towards a New Paradigm of Friction

Analogy is a similarity of non-identical objects (phenomena, processes) in some of their properties. A conclusion made by the analogy principle is as follows. The knowledge gained from a consideration of an object is transferred to a less studied, but similar in essential features, object. Such conclusions are one of the sources of scientific hypotheses.

Analogies in Science: The concepts and the technical tools of modern logic are applied in analysis of systems of scientific knowledge since the first quarter of the 20th century [25]. The analogy principle is one of the logical methods of implementation of the rules and procedures of the study. A usual object's new features search pattern by the analogy principle is in construction of the following logical chain:

– object A has attributes a, b, c, d;

– the object of interest B, similar to A, has attributes a, b, c;

– hence, there is a probability that attribute d is peculiar to object B.

At the stage of science discovery, the analogy occasionally replaced the system observations and experiments. Logical conclusions were often based on the similarity of the external secondary attributes of the studied objects. However, in many cases, analogies made it possible to take a correct first step in choosing a conceptual approach for the construction of a scientific model or theory. For example, the Dutch physicist, H. Huygens, based on the analogy between the properties of light and sound, came (1678) to the conclusion about the wave nature of light. Developing these ideas, Maxwell built (1864) the theory of electromagnetism, which still remains an example of elegancy in physical science. The above-mentioned Garkunov-Polyakov principle of matching the evolutionary paths that enriched the tribology technique is based on analogy.

With the development of scientific theory, the analogies evidence base loses its value. The matches may be the result of accidental similarity of attributes and the reliability of the conclusions based on analogies can be very low. Note the analogy of being (*analogia entis*) – one of the principles of Catholicism justifying the possibility of knowing the existence of God from the existence of the world created by him. Nonetheless, the analogy principle in science has long been serving as a tool for understanding the essence of the problem and choosing the direction to solve it.

To increase the probability of the approximation of scientific findings based on analogies to the truth, the following guidelines are observed [27]:

1) the analogy should be based on the essential attributes and on the greatest possible number of similar properties of the objects to be compared;

2) the relation of the attribute about which the conclusion is made with the common attributes found in the objects should be as close as possible;

3) the analogy should not lead to the conclusion about the similarity of objects in all the attributes;

4) conclusion by analogy should be complemented by a study of the differences between the objects and by a proof that these differences cannot be a reason to reject the conclusions of this analogy.

Analogy between Situations in Physics and Tribology: The analysis of a stalemate situation in modern tribology resembles the dead end, for classical mechanics at the turn of the 19th–20th centuries after the experimental discovery of the fact that the speed of light is independent of the motion of its source. The contradiction between the principle of motion relativity put forward by Galileo and new experimental data on the spatial and temporal properties of physical processes seemed insurmountable.

Galileo's principle of relativity states that in all inertial frames of reference (the law of inertia is just in them) any mechanical process occurs identically (with the same initial conditions), i.e. they are absolutely equal. This corresponds to Newton's corpuscular theory of light, according to which the emission of corpuscles is a mechanical process. The source of light that moves with velocity v emits corpuscles that fly with velocity $c \pm v$, where c is the velocity of the corpuscles in the stationary system and the $\pm$ sign depends on whether the directions of the velocity vectors of light and its source are same or opposite. This conclusion, being natural for classical mechanics, was contrary to the results of fine experiments of a new fundamental physical doctrine—the theory of relativity that originated at the turn of 19th–20th centuries. One of the fundamental conclusions of the relativity theory is the independence of the speed of light from the motion of its source.

This is fairly obvious from the position of wave physics. Wave propagation in any medium is determined by its properties and has nothing to do with the movement of their source, then this medium may be associated with a 'dedicated' system of reference. But this contradicts one of the fundamental principles of classical physics – the Galileo's principle of relativity.

Thus, at the turn of the centuries, it was discovered in physics that the two established facts – the constancy of the speed of light and the principle of motion relativity – are mutually exclusive. A similar situation exists in tribology, where two self-contained friction concepts coexist, based on fundamentally different mechanisms of contact interaction. One may argue that the concept of mechanical cohesion of microroughnesses in sliding contradicts the concept of the atomic-molecular interaction of the solids in contact.

Einstein's Approach to Resolving the Crisis in Physics: There is a famous Greek legend about the Gordian node that was tied in such a way that no one could unravel it. Only Alexander the Great found a solution—he cut the knot with his own sword. The problem that appeared in physics turned into something similar to the puzzle of the Gordian knot, so its quick and brave solution needed a person as significant in science as was Alexander the Great in history. Such a person was the great German physicist, Albert Einstein who was recognized in the 20th century as a synonym for a genius. Einstein's approach to the paradox arisen in physics was ideologically similar to the act of cutting the Gordian knot by Alexander. Einstein chose a simple, yet unexpected, decision. He argued that if the two opposing positions—independence of the light speed from the movement of the source and the principle of motion relativity—are direct generalizations of experimental data, then they should not be in logical contradiction and should be postulated as the basic physical principles—the principle of relativity and the principle of the constancy of the speed of light. They came into history as postulates of the relativity theory (a postulate is a position in a scientific theory that is non-self-evident but taken as the original without proof [25]). Their formulation for the first time was given in Einstein's famous article *On the Electrodynamics of Moving Bodies* published in 1905:

Postulate 1 (Einstein's principle of relativity): The laws by which the states of physical systems undergo change are not affected, whether these changes of state be referred to the one or the other of two systems of coordinates in uniform translator motion;

Postulate 2 (Einstein's principle of invariance): As measured by any inertial frame of reference, light is always propagated in empty space with a definite velocity that is independent of the state of motion of the emitting body.

Einstein argued that, when a new theory is being built, the positions taken 'on faith' are inevitable. They serve as the initial principles of the theory that, as they develop, move from the category of axioms and postulates to the category of proofs. Such was they the destiny of Bohr's postulate which is famous in atomic physics. Analyzing the results of E. Rutherford's experiments which formed the basis of the nuclear model of an atom, he came to the need to assume the existence of stable states of atoms. The development of atomic theory and the appearance of mathematical formalism of quantum mechanics made Bohr's postulates as proved theorems.

Scheme of Approach to the Development of Tribology Paradigm: The Einstein's approach to resolving the paradox of classical physics forms the basis of the algorithm that enables elimination of the contradictions that arose in tribology. We repeatedly paid attention to the existence of similarities in the states of these scientific disciplines corresponding to different historical epochs.

If we follow Einstein (not bad 'guiding star'), it must be borne in mind that all the theoretical principles of tribology are based on experimentally discovered or proven by age-old practice facts, the veracity of which is difficult to doubt. Therefore, they should not be in logical contradiction to each other. Hence, a tribounit manifests itself primarily as a mechanical system or as a physical object where atomic-molecular forces dominate, depending on the experimental technique. This ideology is fundamentally different from the postulates of the molecular-mechanical theory of friction [15]. The latter declares the existence of the atomic-molecular component of the friction force (Fig. 1.1), but all of its equations and methodology of describing the frictional interaction have attributes of the classical mechanistic theory. The approach to the construction of friction paradigm based on considering the interaction of the tribosystem and the measurement system recognizes the correctness of conclusions, drawn by B. Deryagin and G. Yepifanov, which rejected the I. Kragelsky's theory.

The tribosystem, the operation of which realizes multiple, both discovered and still unknown friction, mechanisms, shows the nature of these mechanisms in the measurement of its parameters. Taking this postulate as a basis, we may approach the development of an original paradigm of friction as one that:

1) would consistently include all previous tribological groundwork;

2) would show the maximum respect for the great heritage of theory and practice of friction;

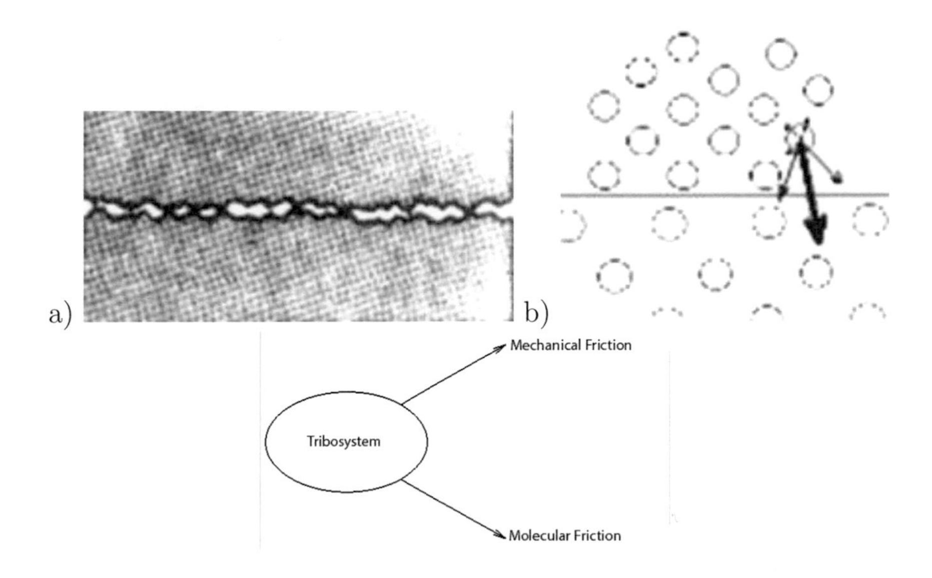

Figure 1.1: (a) mechanical friction; (b) molecular friction; bottom – general model

3) would correspond the requirements of the physical model of a natural phenomenon as formulated by S. Hawking [27].

This brief argument passed to the rank of a separate section in order to, firstly, emphasize the continuity between the ideas of Einstein and the structure of the new paradigm of tribology and, secondly, introduce the concept of disclosure of the inner nature of friction. It is based on the postulate that considers the measurement as a physical effect of a tribosystem.

Chapter 2

Quantum Algorithm for Solving the Fundamental Paradox of Tribology

Based on information given in the first chapter and using the logical constructs of Einstein, we assume that the diversity of properties exhibited by a tribounit in experiments is a reflection of its complex inner essence, each face of which emerges in specific measurements. A question arises whether there is such a theory that would allow us to represent a physical object whose properties are entirely determined by the act of discovery. It is well known that such a theory does exist, and is the most fundamental concept that describes the objective reality. It is quantum mechanics [28].

Quantum mechanics was built on completely different principles than the classical science of Newton and Maxwell. For example, one of the founders of new physics, the outstanding German theorist, Werner Heisenberg, proposed the use of only physically observable quantities in quantum theory. It is very well written by another well-known German physicist, F. Kaempffer: "Quantum mechanics aims to give a description of the physical reality that corresponds to the deliberate exclusion of all the concepts and values not detected in the experiment from the theory" [28]. Exactly this approach to the description of a physical object from the standpoint of quantum theory gives hope for the effectiveness of its staff to study the most general laws of tribological conjunction [28].

2.1 Continuity of the Instrument – Object Relation (Bohr's Measurement Concept)

The great Niels Bohr paid special attention to the theory of physical quantities measurement. He considered the measurement not as a procedure of comparison of the data obtained with some standard but as a physical phenomenon, which consists in the relation of the object (quantum object) and the monitoring tool (the instrument that records the object characteristics).

In Bohr's interpretation, an instrument is a means of knowledge of objective reality that is acceptable for observation, registration and description of various types of measurements. An instrument in quantum mechanics is a specific mediator between the researcher and the object studied by him. The measurements carried out with the help of this tool concern the definition of the object's characteristics that pertain to researchers' interest, which is classically reduced to a comparison of the parameters measured with the standard.

An important factor in the quantum-mechanical measurement procedure is that any act of measurement affects (perturbs) the physical system studied. Scientists fundamentally miss the subtle, gentle agents, with the help of which it would be possible to study the object of interest without touching it. In classical physics, it is permitted to consider the research cycle as a 'contemplation at a distance', which, of course, does not affect the nature of the object of study. However, from the point of view of quantum mechanics, this 'contemplation' is an act of the object interaction with a group of photons that, being scattered on it, impose a part of their energy and the momentum to it, which is identical to the effect of the measurement procedure on the object of study.

For this reason, it must be recognized that between the object of study and the instrument used for conducting the study, there are close links that transform the objects of our scientific research into a single and indivisible instrument—object system (Fig. 2.1) [28].

The existence of an inseparable connection between the instrument and the object of study leads to a number of consequences that determine the specific difference of quantum mechanics from its classic 'forerunner'.

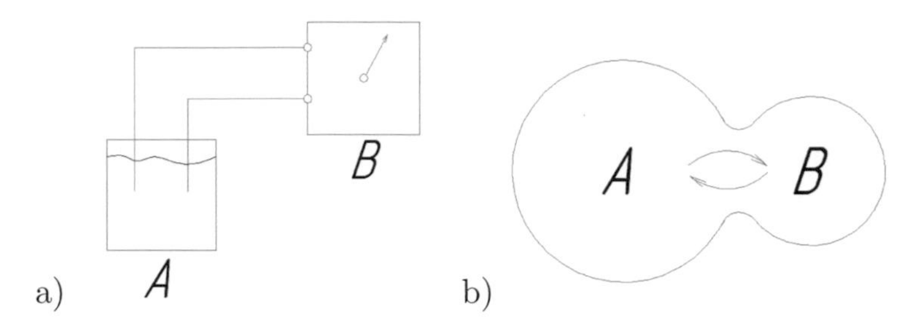

Figure 2.1: Divisibility (separability) of the classical physical system (a) by the object of study A and the instrument B and quantum indivisibility (b) of the object A and the instrument B

Indeed, it follows from the example of the quantum mechanism of remote 'contemplation' that the least relation between the object and the instrument is equal to one quantum of light—a photon. Therefore, the minimum possible relation between the instrument and the object cannot be made arbitrarily small, since it has a flexible minimum equal to $h\nu$, where h is the Planck constant. Then the interaction energy in the ΔW system cannot exceed $h\nu$. Let us express the frequency ν as Δt^{-1}, where t is an arbitrary period of time, then

$$\Delta W \geq h\Delta t^{-1}, \text{ or } \Delta W \Delta t \geq h \tag{2.1}$$

$$\Delta W = \Delta A \tag{2.2}$$

where ΔA is the work, which, in turn, may be written as the product of force and the Δx coordinate:

$$\Delta F \Delta t \Delta x \geq h \tag{2.3}$$

where $\Delta F \Delta t$ is the change in the momentum of force, which, in accordance with Newton's second law, is a change in the momentum of the body Δp. Then

$$\Delta p \Delta x \geq h \tag{2.4}$$

In quantum mechanics, another notation for the Planck constant $\hbar = h/2\pi$ is mostly used, so the relations (2.1) and (2.4) will also be recorded by the $\hbar$ constant.

The expressions (2.1) and (2.4) were obtained by the outstanding German physicist, W. Heisenberg and called the uncertainty principle, which is one of the basic laws of quantum mechanics. The uncertainty principle implies that any physical system cannot be in the state when its coordinate and momentum, as well as time and energy definitely take exact values at the same time. In fact, the uncertainty principle destroyed the concept of determinism, under which there should be a set of laws that allow predicting absolutely everything that might happen in the universe, provided we know the full description of its state at any point of time. At the same time, the uncertainty principle implies the impossibility to know the exact values of certain physical quantities, since an accurate definition of one physical quantity immediately leads to infinite uncertainties in the value of the other. Besides, we must acknowledge that there had been a wreck of principal ideas in classical science that defended the full independence of the object of study from the researcher. Of course, this does not make the studies unscientific because of the 'imaginary subjectivity', since, in accordance with the quantum paradigm, the researcher is 'embedded' in Nature, making a single whole with it. There is even a philosophical direction that fits many doctrines of materialism and idealism in philosophy on the basis of quantum theory [29].

The inextricable connection between the quantum-mechanical object and the instrument imposes specific requirements for the mathematical tools used in quantum mechanics. Physical quantities in quantum mechanics appear as a result of interaction between the instrument and the object based on the fact that "... in the field of quantum phenomena, the exact accounting of the back action of the object on the measuring tool is impossible"; to a large extent it is the action of the instrument itself on the object [30]. In quantum theory, physical quantities are represented by special mathematical parameters that characterize the impact of one value on another and called operators.

The realities of specific physical situations impose restrictions on the form of the operators. Below we present in axiomatic form the basic information from operator algebra necessary for the calculations.

Quantum theory uses linear self-conjugate or Hermitian operators [31].

Definition 2.1. An operator $\hat{A}$ is called linear if the following rule is observed: $\hat{A}(\alpha f + \beta g) = \alpha \hat{A} f + \beta \hat{A} g$, where α, β are constants; f, g are certain functions.

Definition 2.2. An operator $\hat{A}$ is called self-adjoint or Hermitian if its adjoint operator $\hat{A}^*$ or $\hat{A}^+$ ("+" is the Hermitian conjugation sign) is equal to the transposed operator

$$\bar{\hat{A}} = \hat{A}^* = \hat{A}^+$$

Definition 2.3. The sum (difference) of the linear operators $\hat{A}$ and $\hat{B}$ is also a linear operator $\hat{C} = \hat{A} \pm \hat{B}$, whose action on a certain function f is equivalent to the sum (difference) of these operators:

$$\hat{C}f = \left(\hat{A} \pm \hat{B}\right)f = \hat{A}f \pm \hat{B}f$$

Corollary 2.1. *(from Def. 2.3)* For linear operators the commutative and associative law of addition are valid:

$$\hat{A} + \hat{B} = \hat{B} + \hat{A};\ \hat{A} + \left(\hat{B} + \hat{C}\right) = \left(\hat{A} + \hat{B}\right) + \hat{C}$$

Definition 2.4. Multiplying the operator by a number is equivalent to multiplying the result of this operator by that number:

$$\left(a\hat{A}\right)f = a\left(\hat{A}f\right)$$

Definition 2.5. The product of two operators has associative and distributive properties:

$$\left(\hat{A}\hat{B}\right)f = \hat{A}\left(\hat{B}f\right),\ \hat{A}\left(\hat{B} + \hat{C}\right) = \hat{A}\hat{B} + \hat{A}\hat{C}$$

Corollary 2.2. *(from Def. 2.4 and 2.5)* The degree of operators is defined as an action equivalent to sequential operation of the product of base operators and that functions as many times as specified by the exponent: $\underbrace{\hat{A}^n f = \hat{A}\left[\hat{A}\ldots\left(\hat{A}f\right)\right]}_{n}$, accordingly, $\hat{A}^{n+m} = \hat{A}^n \cdot \hat{A}^m$

One may introduce the so-called inverse operator $\hat{A}^{-1}$, whose multiplication by the operator $\hat{A}$ gives a unit operator

$$\hat{1}\text{: } \hat{A}^{-1}\hat{A} = \hat{1}, \text{ i.e. } \hat{1} \cdot f = f$$

Definition 2.6. An operator is called *unitary* U if the condition

$$\hat{U}^+\,\hat{U} = \hat{1} \text{ is satisfied, i.e. } \hat{U}^+ = \hat{U}^{-1}$$

Definition 2.7. Generally, in operator algebra, the equality $\hat{A}\hat{B} = \hat{B}\hat{A}$ is not held, i.e. if the value $\hat{C} = \left[\hat{A}\hat{B}\right] = \hat{A}\hat{B} - \hat{B}\hat{A}$, called a communicator, is introduced, then $\hat{C}$ is generally different from zero.

Corollary 2.3. *(from Def. 2.7)* $\left[\hat{A}\hat{B}\right] = -\left[\hat{B}\hat{A}\right]$, $\hat{A}\hat{A}^{-1} = 0$.

Definition 2.8. An *operators function* is a certain analytically determined value $F(x)$ with an operator $\hat{A}$ is also determined; by analogy with the Taylor series for this function, $F\left(\hat{A}\right)$ is defined as $F\left(\hat{A}\right) = \sum_{n=0}^{\infty} \frac{1}{n!} F^{(n)}(0)\hat{A}^n$ or, for some operators,

$$F\left(\hat{A}, \hat{B}\right) = \sum_{n=0}^{\infty} \sum_{m=0}^{\infty} \frac{1}{n!} \frac{1}{m!} F^{(nm)}(0,0)\hat{A}^n \hat{B}^m$$

Definition 2.9. The same relationship should be kept between operators as between classical quantities.

We must separately dwell on commutators. In quantum mechanics, if an object has the properties described by non-commuting operators $\hat{A}$ and $\hat{B}$ ($\left[\hat{A}\hat{B} \neq 0\right]$), it means the impossibility of the existence of such a state that the object might simultaneously have the properties described by the operators $\hat{A}$ and $\hat{B}$. Therefore, in terms of Bohr's views on measurements, this should ensure that the properties A and B are characterized by the relation of the object of study to various instruments playing the role of specific frames of reference [32], so that if the operators $\hat{A}$ and $\hat{B}$ do not commute, one cannot speak about non-existence of respective properties, regardless of the measurements. If after the measurement of the value A, the non-commuting value B is measured, then A is measured again and there may be a different result. Similar logic complexity prompted N. Bohr to formulate one of the most fundamental laws of quantum mechanics – the principle of complementarity.

Definition 2.10. Obtaining experimental data on some physical parameters that describe the quantum system studied inevitably involves the loss of information about some other complementary parameters.

Based on the two fundamental principles of quantum mechanics—the uncertainty and the complementarity, that result from the continuity of the instrument—object system, consider the impact of data loss on the accuracy of measurements.

Let $\hat{A}$ be the operator that describes a certain parameter; let $\bar{A}$ be the value of that parameter registered by the instrument and ψ be a function characterizing the object of study. Then the well-known, for all the experimenters, measurement variance $\bar{A}^2$ is (we shall explain the sense of absolutely all of the variables used later, for now we just write the expressions for the variance using the notation introduced in quantum mechanics by the outstanding British physicist P. Dirac)

$$\Delta\bar{A}^2 = \langle\psi|\left(\hat{A} - \bar{A}\right)|\psi\rangle \tag{2.5}$$

It is interesting that the variance parameter $\bar{A}^2$ is itself represented, to some extent, by a measure of uncertainty of the values of A. The very same variance of non-commuting physical values, e.g. from the uncertainty principle 2.4, may be a momentum represented by the operator $\hat{p}$ and the coordinate with the operator $\hat{x}$, then their commutator may be written in the form

$$\left[\hat{p}\hat{x}\right] = i\hat{A} \tag{2.6}$$

where i is the imaginary unit $\sqrt{-1}$.

Let us introduce an operator L:

$$\hat{L} = \hat{x} + i\gamma\hat{p} - (\bar{x} + i\gamma\bar{p}) \tag{2.7}$$

where γ is a certain real parameter; $\bar{x}$ and $\bar{p}$ are the values of coordinate and momentum measured in the experiment.

We find the product $\hat{L}^+\ \hat{L}$ (the physical sense of this operation will also be explained later):

$$\begin{aligned} &\langle\psi|\left[\left(\hat{x} - \bar{x}\right) - i\gamma\left(\hat{p} - \bar{p}\right)\right]\left[\left(\hat{x} - \bar{x}\right) + i\gamma\left(\hat{p} - \bar{p}\right)\right]|\psi\rangle \\ &\quad = \langle\psi|(\Delta\bar{x})^2 + \gamma^2(\bar{p})^2 + i\gamma\left[\hat{p}\hat{x}\right]|\psi\rangle \geq 0 \end{aligned} \tag{2.8}$$

The expression (2.8) implies:

$$\Delta\bar{x}^2 + \gamma^2\Delta\bar{p}^2 - \gamma^2\Delta\bar{A} \geq 0 \tag{2.9}$$

Equation 2.9 is valid for any value of the γ parameter. The discriminant of the quadratic trinomial (2.9) must be positive, i.e.

$$\bar{A}^2 - 4\Delta\bar{p}^2\Delta\bar{x}^2 \leq 0, \text{ or } \Delta\bar{p}^2\Delta\bar{x}^2 \geq \frac{\bar{A}^2}{4} \tag{2.10}$$

In fact, the expression (2.10) is a slightly different notation form of Heisenberg's uncertainty principle. It is evident that the more accurately we measure one of the physical parameters, the greater variance comes with the information on the second value, which is equivalent to the data loss about the system as a whole. Thus, we have a contradiction between the need to improve the accuracy of experiments and the inevitability to strengthen the 'scatter' in determining the values of physical quantities. Tribologists-experimenters (not only) know the fact of the growing of random errors after using a more accurate analytical equipment. Often the data obtained from extra-new equipment do not refute the results got in the 'old-fashioned' way or vice versa, the data obtained with the most advanced equipment are statistically unreliable and the scientific community turns to the use of the old analytical base. Therefore, concurrently with the experimenter's desire to equip the laboratory facilities with technical innovations, the clear understanding of the specifics of the studied processes is still of primary importance, including the causes of data loss declared by the principle of complementarity.

The nature of partial data loss in the measurements is shown in the logical paradox of 'both alive and dead cat' suggested by one of the classics of quantum mechanics, Erwin Schrödinger.

E. Schrödinger

In general, quantum mechanics enables three possible states of the cat: the cat is alive, the cat is dead, the cat is both alive and dead at the same time (neither dying nor reanimated).

It is clear that experimentally only the two first states may be registered, while it is likely that the third state cannot be detected. This contradiction is a good illustration of the difference between the latent structure of a quantum system and its realizations represented in concrete acts of measurement [28].

One of the most popular explanations of the 'cat paradox' is the idea of the reduction of the wave function. The wave function is exactly the function f, g, ψ from Def. 2.1–2.10 and the Eq. (2.5) and (2.8), which contain virtually all the information about the physical system. In the measurement, a part of the information is lost. This is because all the links in the quantum world of the macroscopic reality begin and end at the moment of reduction (collapse of the wave function [28]). The act itself takes place in the process of physical impact of the instrument on the object, i.e. in the measurement, since, as N. Bohr supposed that the classic world with its determinism, essentially begins with the instrument. Therefore, the meaningful experiences of Schrödinger with the cat can give only a purely classical result: the cat is either alive or dead (Fig. 2.2).

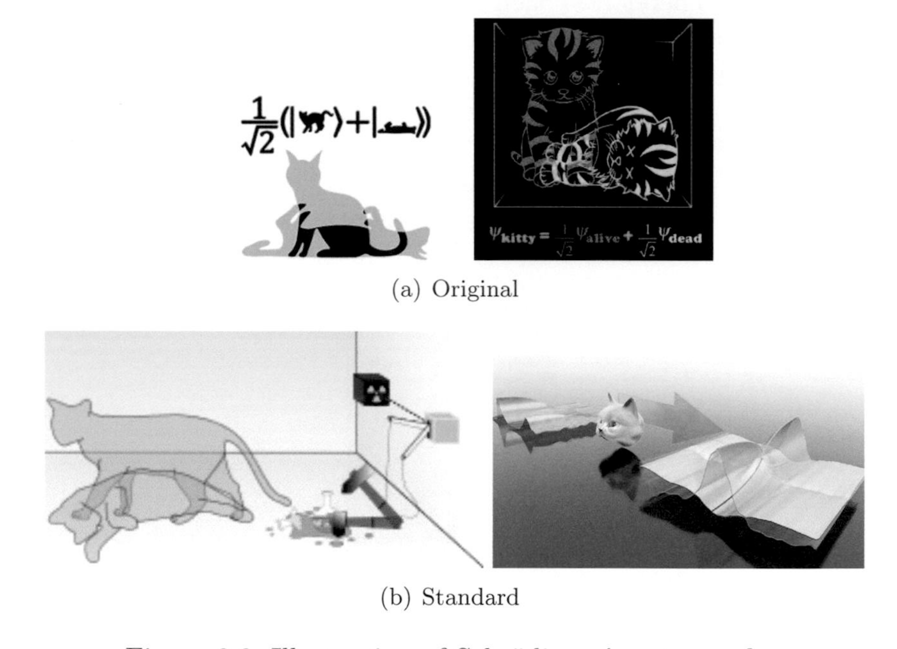

(a) Original

(b) Standard

Figure 2.2: Illustration of Schrödinger's cat paradox

Leading American physicist, Roger Penrose, identified the reduction of the wave function as follows: "As is well known, all the physically interpreted information about quantum-mechanical objects is embedded in their wave functions... The impact of any measuring instrument instantly 'collapses' the wave vector, i.e. there is a reduction (collapse) in the wave function. The Schrödinger equation does not consider such possibilities, therefore Bohr explained the reduction by the fact that during the interaction with the instrument, the physical system transits to the classical form and naturally cannot be described by the Schrödinger equation" [30]. Based on the above, it might seem that the procedure for the reduction of the wave function is a formal mathematical operation whose physical meaning raises doubts of even a number of major specialists in the field of quantum physics. In this sense, very interesting is the discussion of the measurement procedure problem proposed by the American physicist Hugh Everett in 1957.

Everett's model eliminated the internal logical inconsistency of quantum mechanics. In his conception, Everett turned to the analysis of one of the basic principles of quantum theory, under which any impact on the quantum object from a microscopic and therefore classical instrument "immediately causes a collapse of properly wave function that describes the given object" [28].

Thus, in H. Everett's concept, the problem of choice correctness in the measurement result substitutes by the solution of the problem "where the instrument is localized that showed the measurement result obtained" (i.e. in which of the sub-worlds). More recent interpretations of Everett's measurement models considered the behavior of the wave function as a superposition whose members characterize the individual 'sub-worlds'. The terms of that superposition oscillate in time with a certain phase shift with respect to one another. According to the famous German theoretical physicist, D. Zech, the phase shift makes it possible to explain the 'invisibility' of those 'sub-worlds' for one another. Another well-known physicist, R. Feynman's teacher, J. Wheeler came to the conclusion that Everett's multiverse is not just a funny math casus, but it really exists [33].

In the Everett's concept, whenever there is an interaction between a quantum system and a classical instrument or another quantum system, the wave function (sometimes called the wave function of the universe) 'splits', giving rise to a new set of 'histories' *(term adopted from the theory of relativity – author's)*. The choice of one of the parallel worlds may even be the function of the observer's consciousness [33], resulting in an entirely different relational scheme between quantum and classical realities, where the latter only arises after the observer's consciousness chooses one of the parallel worlds. The optimal worlds continue their independent existence. Under one of the 'scenarios' written in Everett's concept, the number of 'universes' propagated by the above laws achieves a fantastically huge number of 10^{60}.

In accordance with the world structure of the Everett's model (Fig. 2.3), the classical world is what we observe and so it is seen as the objective reality. The quantum world exists for us in the form of an abstract image, far from practice, which, nevertheless, enables predicting this very classical reality, although probabilistically. The quantum world exists objectively, but under Everett's interpretation, it stays in the form of a big number of parallel classical world-**plank**tons (named M. Planck) [33].

Hugh Everett

Figure 2.3: Everett's worlds

2.2 The Einstein – Podolsky – Rosen Paradox and the Problem of 'Incompleteness' of Quantum Equations

The loss of information about objects, the researches on quantum measurements results, the destruction of classical determinism because of the uncertainty principle and, consequently, the only alternative made many eminent scientists to carefully apply or criticize the basic principles of quantum mechanics. Among the scientists who questioned the basic positions of this theory was the great Albert Einstein whose work largely predetermined the rise of quantum physics. At different times, to justify the disadvantages of quantum theory, he proposed a variety of situations that, showed the internal contradictions of the new physical concept.

The most famous was one of those tasks that was formulated by Einstein and his followers Boris Podolsky and Nathan Rosen, published in *Physical Review* in 1935 and called the EPR paradox (abbreviated by the first letters of the authors' names). In this publication, the authors reflected Einstein's view of the physical reality and put a mental problem that, in their opinion, showed the incompleteness of description of the objective reality by quantum mechanics.

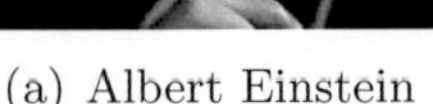

(a) Albert Einstein (b) Boris Podolsky (c) Nathan Rosen

Einstein, Podolsky and Rosen wrote: "Any serious consideration of a physical theory must take into account the distinction between the objective reality, which is independent of any theory, and the physical concepts with which the theory operates. These concepts are intended to correspond with the objective reality and by means of these concepts, we picture this reality to ourselves... . The correctness of the theory is judged by the degree of agreement between the conclusions of the theory and human experience. This experience, which alone enables us to make inferences about reality, in physics takes the form of experiment and measurement... . Whatever the meaning assigned to the term *complete*, the following requirement for a complete theory seems to be a necessary one: *every element of the physical reality must have a counterpart in the physical theory*. We shall call this the condition of completeness... as soon as we are able to decide what are the elements of the physical reality. The elements of the physical reality cannot be determined by a priori philosophical considerations, but must be found by an appeal to results of experiments and measurements. A comprehensive definition of reality is, however, unnecessary for our purpose. We shall be satisfied with the following criterion, which we regard as reasonable. *If, without in any way disturbing a system, we can predict with certainty (i.e. with probability equal to unity) the value of a physical quantity, then there exists an element of physical reality corresponding to this physical quantity...* Regarded not as necessary, but merely as a sufficient condition of reality, this criterion is in agreement with classical as well as quantum-mechanical ideas of reality."

Then Einstein and the co-authors, using the ideas of quantum mechanics, tried to prove that the tools of this theory do not fully describe the picture of the objective reality. They wrote: "...The fundamental concepts of the theory is the concept of *state*, which is supposed to be completely characterized by the wave function ψ, which is a function of the variables chosen to describe the particle's behavior. Corresponding to each physically observable quantity A there is an operator, which may be designated by the same letter. If ψ is an eigenfunction of the operator A, that is, if

$$\psi' = A'_t = a'_t \tag{1}$$

where a is a number,

then the physical quantity A has with certainty the value a whenever the particle is in the state given by ψ *(the original formulae numbering is used here, not related to the continuous numeration in this book—author's)...*

Let, for example,

$$\psi = e^{\frac{2\pi i}{h} p_0 x} \tag{2}$$

where h is Planck's constant; p_0 is some constant number; x is the independent variable.

Since the operator corresponding to the momentum of the particle is

$$p = \frac{h}{2\pi i}\frac{\partial}{\partial x} \tag{3}$$

we obtain

$$\psi' = p\psi = \frac{h}{2\pi i}\frac{\partial \psi}{\partial x} = p_0\psi \tag{4}$$

...in the state given by Eq. (2), the momentum has certainly the value p_0. It thus has meaning to say that the momentum of the particle in the state given by Eq. (2) is real. On the other hand if Eq. (1) does not hold, we can no longer speak of the physical quantity A having a particular value. This is the case, for example, with the coordinate of the particle. The operator corresponding to it, say q, is the operator of multiplication by the independent variable. Thus,

$$q\psi = x\psi \neq a\psi \tag{5}$$

In accordance with quantum mechanics we can only say that the relative probability that a measurement of the coordinate will give is a result lying between a and b and that is

$$P(a,b) = \int_a^b \psi \cdot \psi dx = \int_a^b dx = b - a \tag{6}$$

Since this probability is independent of a, but depends only upon the difference $(b-a)$, we see that all values of the coordinate are equally probable... The usual conclusion from this is that *when the momentum of a particle is known, its coordinate has no physical reality...* From this follows that either 1) *the quantum-mechanical description of reality given by the wave function is not complete* or 2) *when the operators corresponding to two physical quantities do not commute, the two quantities cannot have simultaneous reality*" [34].

To prove their ideas, Einstein et al. proposed considering the following model. They wrote: "...suppose that we have two systems, I and II, which we permit to interact from the time $t = 0$ to $t = T$, after which time we suppose that there is no longer any interaction between the two parts. We suppose further that the states of the two systems before $t = 0$ were known. We can then calculate with the help of Schrödinger's equation the state of the combined system $I + II$ at any subsequent time; in particular, for any $t > T$. Let us designate the corresponding wave function by Ψ. We cannot, however, calculate the state in which either one of the two systems is left after the interaction. This, according to quantum mechanics, can be done only with the help of further measurements, by a process known as the *reduction of the wave packet...* Let $a_1, a_2, a_3, \ldots$ be the eigenvalues of some physical quantity A pertaining to system I and $U_1(x_1), U_2(x_2), U_3(x_3), \ldots$ the corresponding eigenfunctions, where x_1 stands for the variables used to describe the first system. Then Ψ, considered as a function of x_1, can be expressed as

$$\Psi(x_1, x_2) = \sum_{n=1}^{\infty} \Psi_n(x_2) U_n(x_1) \tag{7}$$

where x_2 stands for the variables used to describe the second system. Here $\Psi_n(x_2)$ are to be regarded merely as the coefficients of the expansion of Ψ into a series of orthogonal functions $U_n(x_1)$.

Suppose now that the quantity A is measured and it is found to have the value a_k, it is then concluded that after the measurement the first system is left in the state given by the wave function $U_k(x_1)$ and that the second system is left in the state given by the wave function $\Psi_k(x_2)$. This is the process of reduction of the wave packet: the wave packet given by the infinite series (7) is reduced to a single term $\Psi_\kappa(x_2)U_\kappa(x_1)$.

The set of functions $U_n(x_1)$ is determined by the choice of the physical quantity A. If, instead of this, we had chosen another quantity, say B, having the eigenvalues $b_1, b_2, \ldots$ and eigenfunctions $\theta_1(x_1), \theta_2(x_1)\ldots$ we should have obtained, instead of Eq. (7), the expansion

$$\Psi(x_1, x_2) = \sum_{S=1}^{\infty} \phi_s(x_2)\theta_s(x_1) \tag{8}$$

where $\phi_s(x_2)$ are the new coefficients.

If now the quantity B is measured and is found to have the value b_r, we conclude that after the measurement, the first system is left in the state given by $\theta_r(x_1)$ and the second system is left in the state given by $\theta_r(x_2)$.

We see therefore that, as a consequence of two different measurements performed upon the first system, the second system may be left in states with two different wave functions. On the other hand, since at the time of measurement the two systems no longer interact, no real change can take place in the second system in consequence of anything that may be done to the first system. This is, of course, merely a statement of what is meant by the absence of an interaction between the two systems. Thus, *it is possible to assign two different wave functions... to the same reality...* we assume that Ψ_κ and ϕ_r are indeed eigenfunctions of some non-commuting operators P and Q, corresponding to the eigenvalues p_k and q_r, respectively. Thus, by measuring either A or B, we are in a position to predict with certainty, and without in any way disturbing the second system, either the value of the quantity P... or the value of the quantity Q... we arrived at the conclusion... that two or more physical quantities can be regarded as simultaneous elements of reality only when they can be simultaneously measured or predicted. On this point of view, since either one or the other, but not both simultaneously, of the quantities P and Q can be predicted, they are not simultaneously real. *This makes the reality of P and Q depend upon the process of measurement carried out on the first system, which does not disturb the second system in any way. No reasonable definition of reality could be expected to permit this*" [34].

The answer to such a reasoned and strict criticism of the foundations of quantum theory was the great work by Niels Bohr, published in *Physical Review* in the same year, 1935 [35].

Niels Bohr

Niels Bohr wrote: "...in quantum mechanics, just as in classical mechanics, it is possible under suitable conditions to predict the value of any given variable pertaining to the description of a mechanical system from measurements performed entirely on other systems which previously have been in interaction with the system under investigation. According to their criterion the authors *(Einstein, Podolsky, Rosen—author's)* therefore want to ascribe an element of reality to each of the quantities represented by such variables. Since, moreover, it is a well-known feature of the present formalism of quantum mechanics that it is never possible, in the description of the state of a mechanical system, to attach definite values to both the two canonically conjugate variables, they consequently deem this formalism to be incomplete and express the belief that a more satisfactory theory can be developed... The apparent contradiction in fact discloses only an essential inadequacy of the customary viewpoint of natural philosophy for a rational account of physical phenomena of the type with which we are concerned in quantum mechanics. Indeed the finite interaction between object and measuring agencies conditioned by the very existence of the quantum of action entails – because of the impossibility of controlling the reaction of the object on the measuring instruments if these are to serve their purpose – the necessity of a final renunciation of the classical ideas of causality and a radical revision of our attitude towards the problem of physical reality..." [35].

Next Neils Bohr, from these positions, began to illustrate his own thesis by a mental experiment on a quantum particle passing through various slits that is now often used in quantum physics courses [35]. Having analyzed the different variants of the particle interaction with the diaphragm of the measuring instrument, he wrote: "...in the phenomena concerned, we are not dealing with an incomplete description characterized by the arbitrary picking out of different elements of physical reality at the cost of scarifying other such elements, but with a rational discrimination between essentially different experimental arrangements and procedures which are suited either for an unambiguous use of the idea of space location, or for a legitimate application of the conservation theorem of momentum. Any remaining appearance of arbitrariness concerns merely our freedom of handling the measuring instruments, characteristic of the very idea of experiment. In fact, the renunciation in each experimental arrangement of the one or the other of two aspects of the description of physical phenomena – the combination of which characterizes the method of classical physics, and which therefore, in this sense, may be considered as complementary to one another – depends essentially on the impossibility, in the field of quantum theory, of accurately controlling the reaction of the object on the measuring instruments, i.e. the transfer of momentum in case of position measurements and the displacement in case of momentum measurements. Just in this last respect, any comparison between quantum mechanics and ordinary statistical mechanics, however useful it may be for the formal presentation of the theory, is essentially irrelevant. Indeed we have in each experimental arrangement suited for the study of proper quantum phenomena not merely to do with an ignorance of the value of certain physical quantities, but with the impossibility of defining these quantities in an unambiguous way.

The last remarks apply equally well to the special problem treated by Einstein, Podolsky and Rosen, which has been referred to above, and which does not actually involve any greater intricacies than the simple examples discussed above... in the 'freedom of choice' offered by the last arrangement, just concerned with a discrimination between different experimental procedures which allow the unambiguous use of complementary classical concepts... we now see that the wording of the above-mentioned criterion of physical reality proposed by Einstein, Podolsky and Rosen contains an ambiguity as regards the meaning of the expression 'without in any way disturbing a system'. Of course there is, in a case like that just considered, no question of a mechanical disturbance of the system under investigation during the last critical stage of the measuring procedure. But even at this stage, there is essentially the question of influence on the very conditions which define the possible types of predictions regarding the future behavior of the system. Since these conditions constitute an inherent element of the description of any phenomenon to which the term 'physical reality' can be assigned, we see that the argumentation of the mentioned authors does not justify their conclusion that quantum-mechanical description is essentially incomplete. On the contrary, this description is, as appears from the previous interpretation of measurements, compatible with the finite and uncontrollable interaction between the objects and the measuring instruments in the field of quantum theory. In fact, it is only the mutual exclusion of any two experimental procedures, permitting the unambiguous definition of complementary physical quantities, which provides room for new physical laws, the coexistence of which might at first sight appear irreconcilable with the basic principles of science. It is just this entirely new situation as regards the description of physical phenomena that the notion of complementarity aims at characterizing... In accordance with this situation, there can be no question of any unambiguous interpretation of the symbols of quantum mechanics other than that embodied in the well-known rules which allow to predict the results to be obtained by a given experimental arrangement described in a totally classical way and which have found their general expression through the transformation theorems, already referred to. By securing its proper correspondence with the classical theory, these theorems exclude in particular any imaginable inconsistency in the quantum-mechanical description connected with a change of the place, where the discrimination is made between object and measuring agencies. In fact, it is an obvious consequence of the above argument that in each experimental arrangement and measuring procedure, we have only a free choice of this place within a region where the quantum-mechanical description of the process concerned is effectively equivalent with the classical description" [35].

The controversy of the two most outstanding physicists of the 20th century on the completeness and, accordingly, the correctness of one of the most fundamental concepts of Nature ever created by humanity could not be beyond the field of view of the scientific community. The prominent Soviet theoretical physicist, V.A. Fock, evaluating the positions of Bohr and Einstein, wrote in his introduction [30]: "In quantum mechanics, we are faced with new physical ideals, so different from the usual ideas of the classical theory that their mastering is considerably hard, especially for the minds brought up on classical physics... Quantum mechanics deals with the study of the objective features in Nature, in the sense that its laws are dictated by Nature itself, not by human imagination. However, among the objective concepts there is no concept of the state in the quantum sense. In quantum mechanics, the concept of the state merges with the concept of 'state information', resulting from a certain most accurate experiment. The wave function here does not describe the state in the objective sense, but rather the 'state information'. Einstein showed that, without touching the system, it is possible to give the wave function either form. If we assume, together with Einstein, that the wave function describes the objective state, then, of course, the result will be a paradox in character. Because it is impossible to imagine that the objective system state... changes as a result of whatever operations made on another system that does not interact with it. But even the objective system state cannot change as a result of such operations, the information about the state can change, i.e. the state in the quantum sense... in classical mechanics, any two most accurate experiments provide the same information about the system, and it would then be possible to speak about the system state as something objective, without specifying by what particular experiment the information about it is obtained. This is not the case in quantum mechanics. Here such specifications are fundamentally necessary. Indeed, the Heisenberg's principle shows that different experiences may interfere with each other. Therefore, most accurate experiments may be innumerable: some of them would give the most accurate information about the coordinates, the others about the momenta, etc. Each result of the most accurate experiment corresponds to the wave function in quantum mechanics. It is thus the recording of the information obtained by such an experiment. From this viewpoint, the physical meaning of the dependence of the wave function of time, i.e. the physical sense of the Schrödinger equation, is easy to understand. It allows to use the information related to the initial moment of time for predictions relating to the later moments of time... in quantum mechanics, the distinction between the concepts of 'the state' and 'the most accurate information about the state' erases. This statement is only another formulation of the idea that Bohr means

V.A. Fock

when he speaks about the need for 'a radical revision of our attitude towards the problem of physical reality'. This idea is brilliantly grounded by Bohr in his article".

Next Fock analyzes in detail from Bohr's positions the 'two-component' system, based on the analysis of which, Einstein with his followers, made their conclusions. Fock wrote: "Assume our system consists of two subsystems. The coordinates... of the first subsystem are denoted by x_1, the coordinates of the second one by x_2. Suppose there was the most accurate experiment over the first system that gave us the wave function $\Psi(x_1, x_2)$. Then, although the information on the complete system is the most accurate, the information about the subsystems, generally speaking, will not be accurate. Therefore, even if both subsystems no longer interact, we may not attribute any wave functions, even unknown, to them individually (Einstein apparently thought the opposite). We may only ascribe the statistical operators to them, which are readily expressed as $\Psi(x_1, x_2)$. To get the most accurate information on the individual subsystems, i.e. to determine their wave functions, we must conduct another experiment. The knowledge about the wave function $\Psi(x_1, x_2)$ here makes the simplification that it is enough to conduct this experiment over one of the subsystems. This process is considered by Einstein. From what is said here, it is clear enough that just a misinterpretation of the physical sense of the wave function made him to conclude about the incompleteness of the quantum-mechanical description".

Therefore, if we take the side of the more reasoned position held by both Neils Bohr and V. Fock, we should recognize that while a quantum particle does not interact with any classical object, it has no specific spatial and temporal characteristics, as if being in all the regions of space where the corresponding wave function is non-zero at once. In this manner, for example, one may interpret Feynman's favorite experiment with the passage of a particle through a barrier with two slits and its interference: until the particle is caught by the detector, it is not localized and seems to pass through both slits at once, from those slits the interference is observed [36].

So the information about the quantum system is not necessarily most accurate as viewed by Einstein, Bohr and Fock. The quantum system is not required to have a certain wave function. For example, with a certain probability, momentum of the system is known and there is absolutely no information about its coordinates. In this situation, as John von Neumann suggested, there builds some statistical operator that allows to calculate the probability and the mathematical expectations for all mechanical values corresponding to the available information [36].

After Einstein's and Bohr's articles quoted and discussed in detail above, the results and conclusions of these publications were repeatedly subjected to further analysis, in particular, on what is concerning non-locality, which neither Bohr nor Einstein considered. Gradually, an approach started 'crystallizing' based on the 'Copenhagen interpretation of quantum mechanics' that united supporters and followers of Bohr [28]. Under this approach, quantum systems may not be considered localized before the interaction, so saying that they were scattered over long distances after the measurement, which provided the isolation of quantum particles, makes no sense. Then a question arises about the nature of the interaction of non-local quantum objects. Working in this direction, Fock suggested the 'non-force interaction of quantum objects' [36]. In his comments on Einstein's *Autobiographical Notes,* he writes that Einstein's mistake is exactly in the denial of any interactions except the force ones. The peculiarity of the behavior of the quanta described in the EPR paradox, according to Fock, is a clear evidence of non-force quantum interaction. As an example of another manifestation of similar effect, Fock gives the known quantum correlations that ensure the Pauli exclusion principle.

Finally, a belief arose that the measurement does not necessarily mean a direct interaction between the instrument and the object. An indirect measurement may not fundamentally differ from the direct one.

An original interpretation of non-locality was proposed by David Bohm. As an example, he conducted a mental experiment with a fish in an aquarium. "Imagine an aquarium containing a fish. Imagine also that you are unable to see the aquarium directly and your knowledge about it and what it contains comes from two television cameras, one directed at the aquarium's front and the other directed at its side. As you stare at the two television monitors, you might assume that the fish on each of the screens are separate entities. After all, because the cameras are set at different angles, each of the images will be slightly different. But as you continue to watch the two fish, you will eventually become aware that there is a certain relationship between them. When one turns, the other also makes a slightly different but corresponding turn; when one faces the front, the other always faces towards the side. If you remain unaware of the full scope of the situation, you might even conclude that the fish must be instantaneously communicating with one another, but this is clearly not the case" [23]. In other words, Bohm concluded that elementary particles interact at any distance not because they share incomprehensible and mysterious quanta, but because they are a single entity and their separation is only our illusion.

There is something similar in the statements of John Wheeler, the great teacher of R. Feynman and H. Everett. Feynman recalled an interesting dialogue between them: "I received a telephone call one day at the graduate college at Princeton from Professor Wheeler, in which he said, "Feynman, I know why all electrons have the same charge and the same mass" "Why?" "Because they are all the same electrons!" [37].

It is worth recalling that between electrons there are also correlational 'forces' associated with the Pauli exclusion principle. This approach assumes that any measurement made on one of the correlated particles changes the state of the other and this interaction occurs with an infinite speed. The nature of the correlations does not depend on the distance between the particles. A new term arose to describe this situation in physical systems – quantum entanglement [38].

Most of the articles published on entangled state can be found in the archives of electronic preprints of the Los Alamos National Laboratory, USA, which already emphasizes the high level of interest on this problem that exists at the moment [39]. The term 'entanglement' itself is usually referred to the fact of correlation and interdependency (entanglement) observed in quantum systems.

There are several definitions of quantum entanglement :

Definition 2.11. *Entanglement* is only a special form of quantum correlation that is most convenient to consider by some example. Let us take a system of two particles with a spin of 1/2 (fermions), therefore having the total spin 0 called a singlet state or EPR state; the corresponding correlations are called the EPR correlations. For a system of two spins in the singlet state there is only one quantum state with zero total spin. Therefore, any attempt to 'reconstruct' a part of the system (change the spin direction) instantly leads to a similar reorientation of the spin of the other particle. This effect has no classical or 'quasi-classical' analog [39].

Definition 2.12. *Quantum entanglement* is the state of inseparable integrity and unity. Usually the following definition is given: the entangled state is a state of the composite system that cannot be divided into separate, fully autonomous and independent parts, i.e. it is inseparable (Fig. 2.4) [40].

Definition 2.13. *Entanglement* is a special type of relationship between the component parts of the system that has no analog in classical physics. This relationship is unnatural, inconceivable in terms of classical ideas about reality and looks magical in the literal sense of the word [40].

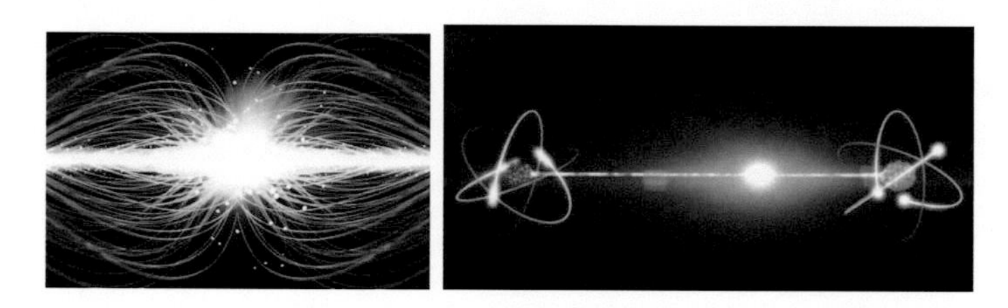

Figure 2.4: Pictorial view of quantum entanglement

The entanglement is not a simple superposition of different quantum states, but such an interlacing that it is impossible to separate one from the other. Interestingly, in the Russian language, the word 'запутанный' ('квантовая запутанность' for 'quantum entanglement') is used in the meaning of confusion and chaos, which is partly symbolic, since such an impression arises from the first acquaintance with this phenomenon. The entanglement is also associated with the presence of hidden connections between the subsystems of physical objects and it is completely inexplicable from the standpoint of the usual field interaction between bodies. The well-known specialist in the field of quantum entanglement S.I. Doronin in his works that cause heated debates compared this phenomenon with telepathy, "when one object directly 'feels' its unity with other bodies, when all the external changes in the object immediately impact on the environment ...the smallest 'wiggling' of any one subsystem is accompanied by simultaneous, harmonized changes in other parts of the system" [38].

Depending on the size of quantum entanglement, the system may consist of individual local parts that loosely connect with one another. In this case, the measure of entanglement of the physical system is close to zero. Accordingly, if the system is a single whole, the measure of entanglement is equal to one. In such a system, being a non-local state, there is a complete lack of separate objects. Therefore, the division may be subject to only the systems in a separable state, although both the absolute entanglement and the complete separability are unlikely to occur in our world. Any object that interacts with its environment is in an entangled state with it. Accordingly, the term of entangled states and quantum correlations may relate not only to quantum systems in the usual sense of the word, but even to complex meso- and macro-bodies composed of many subsystems [39]. In his book, Doronin wrote about quantum correlations in the macroworld: "...while interacting with the environment, we are connected with it by non-local quantum correlations. A question may arise: why then do not we feel these correlations; why do we not feel quantum entanglement? But the fact is that we clearly feel it but do not distinguish it in our mind. Moreover, we have an opportunity to consciously and deliberately change the measure of entanglement..." [38].

The impact on the measure of entanglement may occur through the impact on the surroundings of the physical system. If this impact enhances the 'external rush' on the initially entangled system, then separating it into pieces reduces the measure of entanglement. There is a reverse process, too: the entanglement can be strengthened. The process of system strengthening was called *recoherence* and the process of weakening was called *decoherence*. There is an opinion that decoherence and recoherence are the most fundamental physical processes in the surrounding reality known today in science. The process of decoherence is the basis of all known classical interactions that may be regarded as its consequences.

Entangled states are also divided into pure and mixed ones.

Definition 2.14. *Pure entangled state* is the state of a composite quantum system $Q = A + B + C + \ldots$ whose wave function cannot be represented as a tensor product of the wave function describing its separate parts $\Psi_Q \neq \Psi_A \cdot \Psi_B \cdot \Psi_C \cdot \ldots$

Definition 2.15. *Pure entangled states* are described by density matrices and are to some extent analogous to classical statistical ensembles, since classical correlations may also be observed in them.

Quantum non-locality and entangled states were objects of numerous physical experiments. The first experiments to test the theory of quantum entanglement belong to Professor Anton Zeilinger, University of Vienna. For his experiments, Zeilinger used light quanta – the photons, which, along with electrons, are the most commonly used objects in quantum experiments. An important element of Zeilinger's experiments was their preparatory phase and the basic objects were the photons EPR entangled with one another. For their production, laser radiation was directed to a crystal that had non-linear optical characteristics. The laser radiation was short-term (even on atomic standards) impulses.

At the same time, the visible signals were converted into ultraviolet, which were sent to the crystal splitter, in which a red photon pair was excited. The specific character of the experiment was that both photons were polarized in mutually perpendicular planes, i.e. if oscillations of the electromagnetic field vectors of one photon occurred, e.g. in XOY plane, then those of the other occur in ZOY plane only (Fig. 2.5).

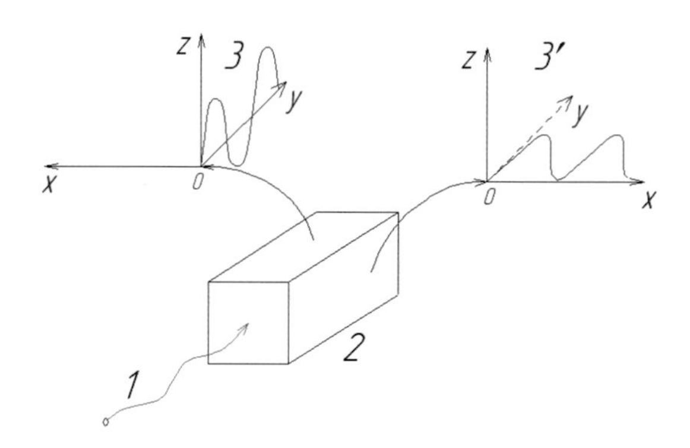

Figure 2.5: Quantum entangled photons in Zeilinger's experiment: (1) the original laser beam; (2) the crystal splitter; (3, 3') EPR entangled photons

So pairs of entangled photons were obtained. The main structural element of Zeilinger's installation was a semitransparent mirror that connected the photons of different pairs with one another. These light particles, depending on the polarization type, either reflected from the mirror surface or penetrated through it. There were four possible options, but in any of them, a quantum entangled pair was formed. The photons properties were now automatically transferred to each other and they became quantum-indistinguishable from their prototypes [28] that were a few meters away from them.

The semitransparent mirror, which is a polarizer, can pass only 'one-way' photons that fall on the detector. Accordingly, the action by rotating the polarizer in the EPR-tied system in the presence of quantum correlations between the scattering photons must be registered by the detector. It is important to note that there is no transfer of particles from one point in space to another in Zeilinger's experiments, since there is already a photon in the detector but only the information about their polarization is transmitted (Fig. 2.6).

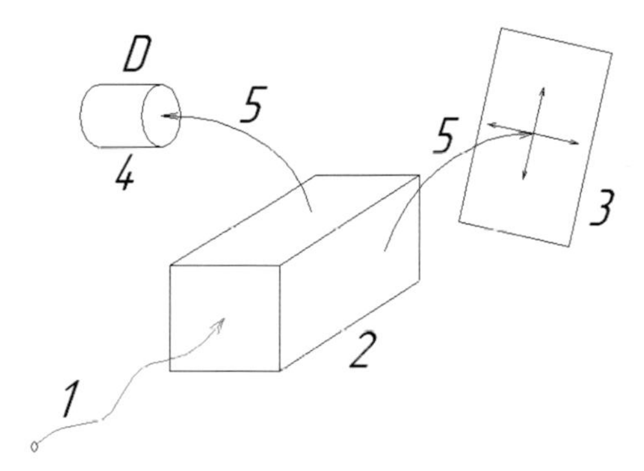

Figure 2.6: Registering the quantum entanglement effect: (1) laser beam; (2) crystal splitter; (3) polarizer; (4) detector; (5) photons in EPR-correlation

The exterior view of A. Zeilinger's experimental stand is shown in Fig. 2.7. After a series of successful experiments, Zeilinger said that he had discovered the effect of quantum teleportation. Now he and his colleagues have learned to teleport up to a hundred elementary particles per hour. French physicists began to conduct experiments with more complex systems and got very encouraging results already on the atom-molecular level.

Digression 2.1. Teleportation is an ultra-fast (faster than light) movement of material objects in space. This idea was repeatedly used in fairy-tales, fantasy literature and science fiction. In this context, the term 'quantum teleportation' sounds challenging and is usually re-

Figure 2.7: Photo of the experimental installation to verify the EPR paradox (a), splitting the laser beam in Zeilinger's experiments (b)

placed by a neutral term, 'quantum entanglement'. The latter term was proposed by Albert Einstein when he tried to prove the absurdity of quantum mechanics [30]. It means that if two particles, whose quantum states are connected, are arbitrarily moved from one another, then changing the state of one of them, leads to a change in the state of the other.

One of the arguments of the opponents of quantum-entangled particle behavior, absurd from the point of view of common sense, was as follows. One may assume that some tiny part of photons involved in the experiment showed the entanglement effect. However, this does not mean at all that the whole ensemble of particles will behave in a similar way. It may well display classical properties, thus shading the quantum effect. If this were the fact, the world would have been even more strange than the world of quantum mechanics. A. Zeilinger's experiments made it possible to measure the polarization of all the photon pairs involved. As expected, they all have demonstrated the effect of quantum entanglement.

Quantum teleportation is simply the use of the term 'teleportation' in the theoretical concept of physics. In most cases it is not associated with mass transfer but simply means that the properties of one particle, e.g. spin, copy to another particle that is remote from the first one. As a synonym for ultra-fast movement, the term 'teleportation' is to a great extent associated with the Schrödinger equation, which tried to describe the path of an electron. It was found that an electron has no path per se. The Schrödinger equation describes the probability of finding the electron at a certain time at a certain point. One may imagine that the electron motion is made by teleportation from one point to another. That is, the electron, being 'caught' in a laboratory, in a moment, may be in another galaxy.

Quantum teleportation is associated with the concepts of the Planck limit. Just as the material has the lowest limit of division – elementary particles, atoms, molecules – so space-time also has its limits of division. The smallest segment of space is called the Planck length and the smallest segment of time is called the Planck time. Thus, the movement of particles in Minkowski space is a teleportation, to which the concept of mathematical continuity is not applied. However, the Planck limits are so small that a series of teleportation jumps are mostly preferred to average formulae described by ordinary motion.

Concepts of the practical application of quantum teleportation are continuously developing.

In his initial experiments, Zeilinger worked with pairs of quantum-entangled photons. Then pairs of entangled electrons and ions were obtained. It soon became possible to create triplets and quartets of entangled photons that have the same wavelengths and which interact with macroscopic objects as a single quantum. The wavelength of this quantum, that consists N photons, decreases N times, allowing for more precise measurements and essentially compress the information [41].

Quantum information transfer over long distances was successfully realized on being implemented for the first time without the use of fiber optics, through the atmosphere. A team of researchers led by Zeilinger was able to transfer the information on the quantum state of a pair of entangled photons between two Canary Islands of La Palma and Tenerife, which are separated by a distance of 143 km. A previous similar experiment, made in China, had the distance of 97 km.

Progress in the field of practical applications of quantum teleportation may radically change, first of all, the usual communication system. In addition to the traditional advantages of quantum data transmission systems – coding density, speed, security – the unique property of quantum teleportation is the ability to implement it even when the location of the receiver and the transmitter are unknown. One can imagine the development of fundamentally new technical systems based on quantum properties of particles: the Internet, quantum computers, non-traumatic medical diagnostic techniques, photographing without the exposure of the light on the object, etc. [42].

The great value of quantum teleportation for science and practice was the ground to nominate Prof. A. Zeilinger for the Nobel Prize in Physics in 2011.

2.3 Information Contents of Quantum Parameters

The unusual behavior of the physical systems that obey the laws of quantum mechanics, which is the connection with the procedure of measuring the physical parameters of the system and the impact of this process on the measured parameters themselves as well as the existence of simultaneously unmeasured physical quantities, makes the quantum-mechanical picture unlike anything known in classical physics. This novelty required to introduce new concepts and to extend the definition of the old ones. The basic concepts of the new physics became the concepts of the system state and wave function. As in all sciences, the definition of the basic term is a very important and complex task, since the contents of the definitions set the boundaries of the theory. As part of the above, let us give definitions of the terms related to the basic concept of a physical quantity.

Definition 2.16. *Physical quantity* is a characteristic of one of the properties of a physical object, qualitatively common for many physical objects but quantitatively individual for each object.

Definition 2.17. *True value* of a physical quantity is the value of a physical quantity that would ideally qualitatively and quantitatively reflect the corresponding physical quality.

Definition 2.18. *Actual value* of a physical quantity is the value of a physical quantity found experimentally and so close to the true value that it can replace it for the current experimental task.

Definition 2.19. *Physical parameter* is the physical quantity considered in the measurement of a given physical quantity as an auxiliary characteristic of that quantity.

Definition 2.20. *Influencing physical quantity* is a physical quantity whose measurement is not provided by the given instrument but which has an impact on the result of the measurement of the physical quantity, for which the instrument is appropriate.

Definition 2.21. *Measured physical quantity* is a physical quantity to be measured or being measured in accordance with the main purpose of the task.

The definitions are taken from the reference monograph [43].

Based on the concept of 'physical quantity' reflected in the Def. 2.15–2.30, we introduce the most important concept of 'state' for quantum mechanics and then the physical quantity that would be its characteristic. In most quantum mechanical courses, the term 'state' is introduced as 'intuitive' and is not defined. Perhaps the first to give a comprehensive formulation of this particular physical quantity was the outstanding English physicist, Paul Dirac. He defines the term 'state' as follows [44]:

Definition 2.22. A *state of a system* may be defined as an undisturbed motion that is restricted by as many conditions or data as are theoretically possible without mutual interference or contradiction. In practice, the conditions could be imposed by a suitable preparation of the system, consisting perhaps in passing it through various kinds of sorting apparatus, such as slits and polarimeters, the system being left undisturbed after the preparation. The word 'state' may be used to mean either the state at one particular time (after the preparation) or the state throughout the entire period of time after the preparation.

Such a cumbersome definition made by Dirac who, according to his colleagues' recollections, was always notable for thoroughness of his formulations and mathematical constructions, shows that, in this case, the semantic load of this term differs from the traditional physical quantities and parameters introduced in Def. 2.15–2.30. The fact is that most physical objects may be reduced to a description in terms familiar in physics: energy, mass, momentum, etc. But in the concept of 'state' we faced a physical quantity whose characteristics cannot be exhausted by only 'traditional real' parameters of classical physics. In Dirac's definition, it is clearly seen that when 'constructing' the concept of 'state', we are faced with a certain qualitative component of our world, which characterizes not so much the individual elements of the physical system as the connections between them, i.e. what is called a structure [45].

In fact, the structure is what gives the organization to the world, thanks to which it does not become a 'chaos' [46]. By studying the structure of an object, we see it as a system [45, 46]. It is clear that relationships within the system should exceed the similar links between the elements of that system and the external environment. With this we take the next logical step that brings us closer to the realization that the system determines the environment to a greater extent than the environment determines the system. The system structure has a diversity, the measure of which is the information [46].

The traditional inertial and energy components of the objective world are mutually convertible. The third information component – the structure – is special. It is related to the other two by the substrate nature, i.e. the structure is always imprinted on the mass-energy level, but the structure itself does not depend on this basis. The latter "...is easy to imagine by the example of a letter: any letter is a number of certain characters deposited onto a substrate (paper, magnetic carrier, etc.), but neither paper itself nor the method of making notes on it defines the meaning of what is written" [46].

The physical quantity introduced to characterize the state of a quantum system should reflect the 'data characteristics' of its description. In 1926, the eminent German physicist, Max Born proposed a hypothesis (now generally accepted) that the state of a quantum system must be described by means of the wave function ψ that we had already met before but did not specify it. According to Born, the wave function has a probabilistic interpretation. He wrote: "Once more an idea of Einstein gave the lead. He had sought to make the duality of particles (light quanta or photons) and waves comprehensible by interpreting the square of the optical wave amplitudes as probability density for the occurrence of photons. This idea could at once be extended to the ψ-function: $|\psi^2|$ to represent the probability density for 'electrons (or other particles)' [47]. In accordance with the Born's concept, a particle is more likely in the location where its wave function is big and less likely where its wave function is small. Because of that, the wave function in quantum mechanics is often called the probability amplitude or simply the amplitude [48]. The emergence of such a concept as a wave function in Born's interpretation declares that the probability is no longer an unawareness but reflects the most fundamental properties of matter.

Max Born's assumption about the nature of wave function allows considering the latter as a carrier of information about the quantum system, the information field characteristics devoid of a definite (in classic view) physical sense. It is known that the measure of quantum entanglement of a physical system is directly related to the information contained in it [40]. This relation allows description of the quantum system in terms of information. For example, the processes of quantum entanglement, strengthening or weakening, may be regarded in terms of information exchange between the physical system and its environment. As for information, it has one of the essential features that distinguish it from the 'normal' physical parameters: that is, it can disappear.

The law of information loss was formulated by the outstanding Russian mathematician, A.A. Lyapunov. To prove his reasoning, he gave the following argument: "...let us assume that we have received the information that is absolutely identical to the once lost one. How can we ascertain this identity? After all, any comparison becames impossible because of the loss. Therefore, it is necessary to recognize the available information as new. That is, in the world where the natural sciences are studied, only one thing can completely disappear – the information" [40].

The law of information (or the structure that carries it) loss is considered the fundamental position in natural sciences, the consequences of which are as global as the consequences from the laws of conservation of energy or conservation of momentum, as we have already managed to make sure when considering the information loss for the state of 'alive-dead cat' of Schrödinger. In quantum theory, any closed system is in a non-local state due to the fact that there is no external environment and no one to carry out the procedure of reduction of the wave function. This is actually 'non-material' state in the sense that it is impossible to describe it in inertial and energy parameters. It is only expressed in terms of quantum information. S.I. Doronin calls this state 'pure information' [40]. In fact, the non-local quantum state is a void where there is no particles nor fields. Our material world, from the viewpoint of quantum theory may be regarded as "the secondary formation within the comprehensive source of quantum reality", which is a manifestation of quantum decoherence of the original quantum information in locally material forms of the material world [40]. That is, there is a birth of 'all out of nothing', as figuratively said the largest modern physicist and cosmologist Stephen Hawking: "...indeed, it [the universe] is literally born out of nothing – not from a vacuum but totally out of nothing because there is nothing beyond the universe" [27].

In this case, the procedure of reduction of the wave function at the measurement is a mathematical technique for calculating the behavior of a quantum system after it had 'interacted' with its 'environment', which 'materialized' one of the possible variants of this interaction. Under the concept of H. Everett, other potential possibilities never disappear and may occur at any time. The decoherence process itself is seen as the folding of initial states space into a smaller space. The current version of reality continues existing for other subsystems that are still in the same initial event space. There is a transition of our object from one level of reality to another. This is not just 'alternative' versions of familiar reality but, according to Doronin, it is 'indeed other levels' – other state spaces with their own spatial metric. All the above is reflected in one of the most fundamental principles of quantum mechanics – the superposition principle [40].

According to the definition given by P. Dirac:

Definition 2.23. if we consider the superposition of two states, A and B, such that there exists an observation which, when made on the system in state A, is certain to lead to one particular result a, and when made on the system in state B is certain to lead to some different result b. The result will be sometimes a and sometimes b, according to a probability law depending on the relative weights of A and B in the superposition process. The superposition principle in quantum mechanics is applied to the states of arbitrary mechanical systems, leading us to assume that there are special relations between these states, such that if the system is entirely in one particular state, then we may assume that it is partly in each of several states. Each newly formed state should be regarded as the result of a superposition of two or more new states. The nature of the relationship between the states of any system arising from the superposition principle is that it cannot be explained in terms of familiar concepts of classical physics, since it is impossible to imagine that the system is partly in one and partly in another state, which is equivalent to saying that the system is entirely in a third state. If the state is formed by the superposition of two other states, it will have the properties in which there are elements of both initial states, to a greater or lesser extent approaching the properties of sometimes one, sometimes another, depending on the coefficient with which the wave functions that describe these states entered into a superposition.

Based on Dirac's definition of the superposition principle, the 'entire quantum reality' may be presented in the form of a multilevel system, in which each level is a separate Everett world, a separate reality with its own metric and energy characteristics. The entire reality, in turn, appears as a superposition of all these levels—the 'worlds' where quantum transitions are possible between, although a direct study of one level from another is impossible. The interaction between the levels—the worlds of the entire quantum system apparently can be carried out only by means of quantum non-local correlations.

Quantum physics does not exclude the possibility that, in addition to decoherence effect, when a 'transformation' of the primary information field of the quantum system is observed, from non-local to local forms that embody the objects of study, the reverse process of recoherence may occur consisting of delocalization of objects and 'dissolving' them in space of the quantum system. The theory of quantum entanglement and associated phenomena of decoherence and recoherence, as well as its main results, are formulated in terms of systems and subsystems. Therefore, we may assume that the consequences of this theoretical model will be valid not only for microparticles but also the meso- and macrosystems produced by them.

Summing up this aspect of quantum mechanics, one may state that one of the consequences of the theory of entangled states is the conclusion that a system in a non-local, superposition state cannot possess the structural elements that may be regarded as local classical bodies. The latter appear in our space only after the completion of decoherence, at which the classical and quantum worlds are virtually indistinguishable, when it is impossible to distinguish the prediction of quantum theory and the conclusions obtained by the tools of classical physics. Perhaps that is why one of the most erudite theorists, von Neumann, wrote: "Strictly speaking, a state is only a theoretical construct, in reality at our disposal are only the results of measurements..." [40]. In terms of quantum superpositions, the 'Schrödinger cat' cannot be in a half-alive and a half-dead state at the same time, which is possible in our world. The cat is alive and dead at the same time, and this is just an example of a non-local quantum state: the cat, being an ordinary physical body, is absent in our classic material world. According to O.O. Feigin, "...It is in a state of a more general type, and the local state is only one particular case, one of the possible options for our cat's being" [28].

2.4 Changes in Physical Parameters of Quantum Systems as a Result of Measurement

The development of new physical theory had its own peculiarities characterized by the well-known Soviet physicist-academician L.I. Mandelstam in his lectures on quantum mechanics: "Classical physics was mostly that the establishment of relations between mathematical values with real things preceded the equations, i.e. the establishment of laws; to find the equations was the main task because the quantities content seemed in advance clear and independent from the laws. Modern theoretical physicists... took another path than classical physics. It happened by itself. Now, first of all, they try to guess the mathematical tools that operate the values, for which or for a part of which it is unknown at all as to what they mean." To some extent, the task ahead is correlated with the words of the eminent Soviet academician, but our approach can also be seen as a return to the 'classics'.

Indeed, we are required to 'guess' the mathematical formalism that describes the transformation of a quantum tribosystem into a measurable image. In this case, both the fundamental meanings of physical parameters and the mathematical relationship between them in the form of quantum-mechanical equations are unknown. The physical object itself is quite unusual and difficult for traditional quantum systems, which affect the mathematical form of the equations and methods used [37]. Nonetheless, based on what was said in Sections 2.1 and 2.3, we may state the following:

a) quantum systems are an indissoluble unity of the classical instrument (measuring the physical quantities) and the quantum object itself;

b) the impact of the instrument on the object is described by Hermitian operators that obey the rules set in Definitions 2.1–2.10;

c) the Hermitian operators are analogs of the physical parameters according to Definition 2.18;

d) the state of a quantum system is defined by a quantity that has no analog in classical physics – the wave function or probability amplitude, which characterizes the information field constituting the fundamental principle of the non-localized (free) quantum system;

e) the wave function may be represented as a superposition, a linear combination of the wave functions that describe the specific (basic) states that can be assumed by the given quantum system;

f) while measuring, a reduction of the wave function occurs, accompanied by the loss of a part of information and the transformation of a purely quantum object into a classic one.

To construct a theoretical (mathematical) model, it is necessary to extend the definition of some terms concerning the wave function and operations with it. The additional conditions imposed on the wave function ψ are associated with the requirement to maintain the physical meaning of the Schrödinger equation [31]:

a) Unambiguity: The motion of a quantum system is expressed by the wave function ψ that must be a unique function of coordinates and time, since otherwise it turns out that for one point of the phase space there are several probabilities of realization of a definite event, which makes no physical sense.

b) Finiteness: The function ψ should everywhere be a finite quantity and vanish at infinity together with its first derivative.

c) Continuity: As a state of a quantum system described by the wave function is similar to a trajectory in classical mechanics, the quantum state of the system in space and time should change continuously; together with the continuity of the wave function, a similar requirement applies to the first derivative of the wave function.

Here are a few definitions that are important for our theoretical constructs [49]:

Definition 2.24. If the conditions defined by the equation

$$\hat{A}\psi = a_n \psi_n \tag{2.11}$$

where $\hat{A}$ is the Hermitian operator of the physical quantity; ψ is the wave function; a_n is a complex number; ψ_n is the eigenvalue of the wave function ψ, is satisfied, then a_n are called *eigenvalues* of operator $\hat{A}$. They are valid for the case of Hermitian operator and (2.11) is called the *eigenvalues equation* .

Definition 2.25. The eigenvalues of Hermitian operators are *orthogonal* with respect to the other eigenvalues of Hermitian operators

$$(a_n - a_m)\delta_{nm} = \Delta a \begin{cases} \neq 0 \text{ if } n = m \\ = 0 \text{ if } n \neq m \end{cases} \tag{2.12}$$

Equation (2.12) implies that if an Hermitian operator $\hat{\xi}$ satisfies the exponential algebraic equation

$$Q\left(\hat{\xi}\right) = \xi^n + a_1\xi^{n-1} + \cdots + a_n \equiv (\xi - \xi_1)(\xi - \xi_2)\ldots(\xi - \xi_n) = 0$$

and if (2.11) is $Q_n(\xi)A = 0$ for this case, then for any case described by vector A there does not exist any $Q_{n'}(\xi) = 0$ with $n' < n$. The consequence is that the operator $\hat{\xi}$ may have only n eigenvalues forming a complete system.

Definition 2.26. The eigenvalues are called *nondegenerate* if each of them, accurate within some constant, corresponds to only one eigenfunction ψ_n; otherwise the eigenstates are degenerate.

Definition 2.27. Even in the case when there is a degeneracy, it is possible and convenient to choose a funcional basis that complies with the requirements of orthogonality (2.12).

Hermitian operators have a property that can be formulated in the form of a quantum-mechanical theorem (lemma), the proof of which is given in detailed courses of quantum mechanics [49], but here we limit ourselves to the following definition:

Lemma 2.1. *If the Hermitian operator $\hat{N}$ has two other Hermitian operators $\hat{L}$ and $\hat{M}$ commuting with it: $\left[\hat{L}\hat{N}\right] = \left[\hat{M}\hat{N}\right] = 0$, while they themselves do not commute, i.e. $\left[\hat{L}\hat{M}\right] \neq 0$, then the eigenvalues of the operator $\hat{N}$ are degenerate.*

The process of measuring quantum quantities requires a definition of another parameter introduced by P. Dirac.

Definition 2.28. An *observable quantity* is a quantity that can be measured; conversely, any observable quantity can be measured at least theoretically [44].

Dirac describes in much details the behavior of quantum quantities and their transformation into the 'observable'. He wrote: "When we make an observation, we measure some dynamical variable. It is obvious physically that the result of such a measurement must always be a real number, so we should expect that any dynamical variable that we can imagine must be a real dynamical variable... We... have to restrict the dynamical variables that we can measure to be real... If the dynamical system is in an eigenstate of a real dynamical variable ξ, belonging to the eigenvalue ξ', then a measurement of ξ will certainly give as result the number ξ'. Conversely, if the system is in a state such that the measurement of a real dynamical variable ξ is certain to give one particular result (instead of giving one or other of several possible results according to a probability law, as is in general the case), then the state is an eigenvalue of ξ and the result of the measurement is the eigenvalue of ξ to which this eigenstate belongs... We can infer that, with the dynamical system in any state, any result of a measurement of a real dynamical variable is one of its eigenvalues. Conversely, every eigenvalue is a possible result of a measurement of the dynamical variable for some state of the system, since it is certainly the result if the state is an eigenstate belonging to this eigenvalue" [44].

Based on clear Dirac's ideas, we may formulate the following statement:

Statement 2.1. The only possible result in measurement of a dynamic quantity represented as a function of coordinates x, momentum p and time t: $F(x, p, t)$ are the eigenvalues of the Hermitian operator $\hat{F}$ corresponding to this function. The quantum-mechanical state of the system is determined by the wave function ψ, which changes in time in accordance with the requirements laid by the Schrödinger equation. So is measured a physical quantity $F(x, p)$ whose result of measurement matches one of the eigenvalues of the Hermitian operator $\hat{F}$, e.g. F_n.

In accordance with Statement 2.1 and the eigenvalues equation (2.11), one may note:

$$\hat{F}g(x) = F_n g_n(x) \tag{2.13}$$

In accordance with the principle of superposition, the wave function ψ describing the quantum system considered may be represented as a linear combination

$$\begin{aligned} \psi &= \sum_n b_n g_n(x) \\ b_n &= \int_x g_n^* \psi dx \end{aligned} \tag{2.14}$$

The linear expansion coefficient characterizes the probability of obtaining the value of F_n as a result of measuring $F(x, p)$. The value b_n characterizes the probability amplitude of this process (we shall discuss it in more detail below), i.e. the probability is proportional to $|b_n|^2$, hence,

$$\sum_n |b_n|^2 = 1 \tag{2.15}$$

The result of the measurement may be considered a mean value of $F(x, p, t)$, which we denote by $\bar{F}$. By definition of the mean value $\bar{F}$, it can be calculated from the ratio [50]

$$\bar{F} = \sum_n |b_n|^2 F_n \tag{2.16}$$

Definition 2.29. The mean value of a Hermitian operator is a real number; the converse is also true: if the mean value of an operator is a real number, then the operator is Hermitian.

Definition 2.30. The mean value $\bar{F}$ of an operator $\hat{F}$ with respect to a wave function ψ may be written as

$$\bar{F} = \frac{\int \psi^* \hat{F} \psi dx}{\int \psi^* \psi dx} \tag{2.17}$$

Expression (2.17) may be written in a different form, using the so-called bra- and ket-vectors introduced into quantum mechanics by P. Dirac who, in his famous course *The Principle of Quantum Mechanics,* wrote: "It is desirable to have a special name for describing the vectors which are connected with the states of a system in quantum mechanics, whether they are in a space of a finite or an infinite number of dimensions. We shall call them ket-vectors and denote a general one of them by a special symbol $\big|\big\rangle$. If we want to specify a particular one of them by a label, A say, we insert it in the middle, thus $\big|A\big\rangle$. Ket-vectors may be multiplied by complex numbers and be added together to give other ket-vectors. Whenever we have a set of vectors in any mathematical theory, we can always set up a second set of vectors, which mathematicians call the dual vectors. The new vectors are, of course, defined only to the extent that their scalar products with the original ket-vectors are given numbers, but this is sufficient for one to be able to build up a mathematical theory about them. We shall call the new vectors bra-vectors and denote a general one of them by the symbol $\big\langle\big|$, the mirror image of the symbol for a ket-vector. If we want to specify a particular one of them by a label, B say, we write it in the middle, thus $\big\langle B\big|$. The scalar product of a bra-vector will be written as $\big\langle B\big|A\big\rangle$, i.e. as a juxtaposition of the symbols for the bra- and ket-vectors, that for the bra-vector being on the left, and the two vertical lines being contracted to one for brevity. We have the rules that any complete bracket expression denotes a number and any incomplete bracket expression denotes a vector, of the bra or ket kind, according to whether it contains the first or second part of the brackets '$\big\langle\big\rangle$'. On account of the one-one correspondence between bra-vectors and ket-vectors, any state of our dynamical system at a particular time may be specified by the direction of a bra-vector just as well as by the direction of a ket-vector. The relationship between a ket-vector and the corresponding bra makes it reasonable to call one of them the conjugate imaginary of the other. To call attention to this distinction, we shall use the words 'conjugate complex' to refer to numbers and other complex quantities which can be split up into real and pure imaginary parts, and the words 'conjugate imaginary' for bra- and ket-vectors, which cannot" [44].

R. Feynman brilliantly explained the physical meaning of Dirac's formalism he was a supporter of. In his lectures on physics, in Quantum Mechanics, Feynman wrote: "... we will use... a shorthand notation – invented by Dirac and generally used in quantum mechanics... We write the probability amplitude this way:

$$\left\langle \text{Particle arrives at } x \middle| \text{particle leaves } s \right\rangle$$

In other words, the two brackets $\left\langle\right\rangle$ are a sign equivalent to 'the amplitude that'; the expression at the *right* of the vertical line always gives the *starting* condition and the one at the *left*, the *final* condition" [48].

Let us rewrite some of our relations in the Dirac's formalism of bra- and ket-vectors. For example, Eq. (2.14) for a coefficient b_n takes the form:

$$b_n = \left\langle g_n \middle| \psi \right\rangle \tag{2.18}$$

and the expression for a mean value (2.17) is rewritten as:

$$\bar{F} = \frac{\left\langle \psi \middle| \bar{F}\psi \right\rangle}{\left\langle \psi \middle| \psi \right\rangle} \tag{2.19}$$

We can accurately obtain the expression (2.19) showing at the same time the technique of working with bra- and ket-vectors. According to the definition, the mean value of any physical quantity is defined by the formula (2.16), in which the square of the coefficient b_n, which is generally a complex number, may be considered a product $b_n^* \cdot b_n$, where b_n^* is a complex conjugate value b_n. Then $\bar{F} = \sum\limits_n |b_n|^2 F_n = \sum\limits_n b_n^* b_n F_n$. In accordance with Def. 2.24, the expression for $\bar{F}$ may be rewritten as:

$$\sum_n b_n^* b_n F_n = \sum_n \sum_s b_s^* F_n b_n \delta_{sn} = \sum_n \sum_s b_s^* F_n b_n \left\langle g_s \middle| g_n \right\rangle$$
$$= \left\langle \sum_s b_s g_s \middle| \sum_n F_n b_n g_n \right\rangle$$

Then, according to the first equation (2.14), the obtained expression is rewritten as

$$\left\langle \sum_s b_s g_s \middle| \sum_n F_n b_n g_n \right\rangle = \left\langle \psi \middle| \hat{F}\psi \right\rangle = \frac{\left\langle \psi \middle| \hat{F}\psi \right\rangle}{\left\langle \psi \middle| \psi \right\rangle}$$

which corresponds to the formula (2.19).

Using the Dirac's formalism, Feynman showed the formal mathematical aspect of an act of quantum measurement. He brought the measurement effect on a quantum system described by the amplitude (wave function) ϕ to a change of the latter to a certain value χ. The impact of the instrument was in a way denoted by j. As a result of a measurement act, inside the 'instrument – object' system a change of the initial quantum object is made, expressed by replacing the amplitude ϕ by amplitude χ; the reverse transition meaning the transformation of the initial object ϕ into χ occurred in a way denoted by i. The action of the instrument is described by the Hermitian operator A [48].

All the above may be represented by vector equations:

$$\begin{gathered}\left\langle \chi \middle| \hat{A} \middle| \phi \right\rangle = \sum_i \sum_j \left\langle \chi \middle| i \right\rangle \left\langle i \middle| \hat{A} \middle| j \right\rangle \left\langle j \middle| \phi \right\rangle \\ \text{or} \\ \left\langle \chi \middle| \psi \right\rangle = \left\langle \chi \middle| \hat{A}\phi \right\rangle \\ \left| \psi \right\rangle = \hat{A}\left| \phi \right\rangle = \sum_i \sum_j \left| i \right\rangle \left\langle i \middle| \hat{A} \middle| j \right\rangle \left\langle j \middle| \phi \right\rangle\end{gathered} \tag{2.20}$$

A more detailed examination of the quantum process of measuring a physical quantity makes us conclude that the problem is reduced to consideration of the same object, but as it were from another frame of reference. In quantum mechanics, this procedure is regulated as a notation of the wave function in different bases. For example, let initial amplitude ϕ be noted in a basis T_j, which from the viewpoint of the principle of superposition (2.14) looks like

$$\begin{gathered}\phi = \sum_j b_j T_j(x) \\ b_j = \left\langle T_j \middle| \phi \right\rangle\end{gathered} \tag{2.21}$$

The function χ, because of the change, 'appears' in another basis S_i, which is noted as

$$\chi = \sum_i c_i S_i(x)$$
$$c_i = \left\langle S_i \middle| \chi \right\rangle \tag{2.22}$$

In accordance with Eq. (2.20), the transition of the quantum system described by the amplitude ϕ from the basis T_j to basis S_i may be represented by vector equation

$$\left\langle S_i \middle| \phi \right\rangle = \sum_j \left\langle S_i \middle| T_j \right\rangle \left\langle T_j \middle| \phi \right\rangle \tag{2.23}$$

When a transition from basis to basis occurs, the following ratios must be met:

$$\sum_i \left| \left\langle \chi \middle| S_i \right\rangle \right|^2 = \sum_j \left| \left\langle \chi \middle| T_j \right\rangle \right|^2$$
$$\left\langle \chi \middle| \phi \right\rangle = \left\langle \phi \middle| \chi \right\rangle^* \tag{2.24}$$

The procedure of rewriting the wave function on another basis is in quantum mechanics similar to the 'transfer' of a physical system to a different phase space. In fact, a measurement as a physical act 'projects' the complex quantum-informational nature onto our familiar four-dimensional world [27, 28].

Chapter 3

Quantum Spaces of Tribosystems

The analysis of the material collected in Chapter 2 allows us to consider the quantum transformations that occur in physical systems as probabilistic space-time changes of some informational substrate. At the same time, our physical-geometric approach makes it possible to represent the act of measuring a quantum physical quantity a multidimensional information field object, as a projection onto the 'plane' of our world – *four-dimensional Minkowski continuum* , in which metric vectors are real values reflecting the specificity and the result of the measurement process. To create an objective physical model of this mapping of a part of a multi-dimensional object on our dimension, it is necessary to make a mathematical model of such multidimensional spaces.

3.1 The Concept of Linear Vector Spaces

Since this is not a book on linear algebra, we omit theoretical calculations and will stop on definitions of the mathematical concepts that will be needed for future quantum-mechanical constructs [51].

Definition 3.1. A collection L of any elements may be called *vector space* over a given number field K if

a) there is a rule that assigns to every two elements a and b of a given set the third element $(a + b)$ of the same set called the sum of the elements a and b;

b) there is another rule that assigns to each element a of a given set and to each number k of a number field K the element ka of set L;

c) both of these rules satisfy the following axioms:

1c. for any elements a, b, c of a set L, the properties of commutativity and associativity are valid;

2c. the set L has a zero element, the addition and subtraction of which with any element a do not change that element;

3c. for any element a of a set L there exists an opposite element, the addition of which with a gives the zero element of the set L;

4c. for any elements a and b of a set L and for any numbers k_1 and k_2 of a field K, the following relations take place:

$$k_1(k_2a) = (k_1k_2)a$$

$$(k_1 + k_2)a = k_1a + k_2a$$

$$k_1(a + b) = k_1a + k_1b$$

d) for any element a of a set L, there is a relation $1 \cdot a \equiv a$, in other words, the multiplication by one does not change the element of the set L.

Definition 3.2. A vector system M of a vector space is considered *linearly independent* if the linear combination of those vectors is equal to zero only if all the coefficients are equal to zero.

Definition 3.3. Any maximal independent system of vectors $e_1, e_2, \ldots, e_n$ of a space L is called a *basis* of that space.

If $e_1, e_2, \ldots, e_n$ is a basis of n-dimensional space L and $e'_1, e'_2, \ldots, e'_n$ is an arbitrary linearly independent vector system of the same space, the system can be supplemented by several vectors of the basis of n-dimensional space L.

Definition 3.4. A *subspace* of a given vector space is a set of vectors of this space with the following properties:

a) if vectors a and b belong to a set L, their sum also belongs to this set;

b) if a vector a belongs to a set L, its multiplication by a certain number k of the number field must also belong to the set L.

Definition 3.5. The *intersection* of two subspaces is itself a subspace; as a consequence, one may argue that any subspace is a self-subspace; the dimension of any subspace must not exceed the dimension of the space.

Definition 3.6. Any vector x of a linear space L can be represented with a basis by a linear combination:

$$x = e_1 x_1 + e_2 x_2 + \cdots + e_n x_n + \ldots$$

where $x_1 \ldots x_n \ldots$ are the coordinates of the vector x;

a) let $e_1, e_2, \ldots, e_n, \ldots$ be the basis of a space L, let a system of independent vectors $e'_1, e'_2, \ldots, e'_m, \ldots$ exist and each last vector be unambiguously represented by a linear combination:

$$e'_1 = e_1 c_{11} + e_2 c_{21} + e_3 c_{31} + \cdots + e_n c_{n1} + \ldots$$

$$e'_2 = e_1 c_{12} + e_2 c_{22} + e_3 c_{32} + \cdots + e_n c_{n2} + \ldots$$

....

$$e'_m = e_1 c_{1m} + e_2 c_{2m} + e_3 c_{3m} + \cdots + e_n c_{nm} + \ldots$$

Then the matrix $c_{nm} = \begin{Vmatrix} c_{11} & c_{21} & \ldots & c_{n1} \\ c_{12} & c_{22} & \ldots & c_{n2} \\ \ldots & \ldots & \ldots & \ldots \\ c_{1m} & c_{2m} & \ldots & c_{nm} \end{Vmatrix}$ is called the *transfer matrix* from the basis e to the basis e'.

Definition 3.7. For the system $e'_1, e'_2, \ldots, e'_m, \ldots,$ to the basis of a space L, it is necessary and sufficient that the transfer matrix C_{nm} be square and that its determinant be non-zero. Therefore, the knowledge of the coordinates of the vector with respect to any of the bases considered is sufficient to determine them with respect to another basis, if the transition matrix C_{nm} is known.

Definition 3.8. A vector system containing more than one element is *linearly dependent* if and only at least one of its vectors is a linear combination of the others; the corollary is that if a vector system contains the zero vector, it is certainly linear dependent.

Definition 3.9. Two vector systems are *equivalent* if any vector of one system may be represented as a linear combination of vectors of the other system; hence, any vector system is equivalent to itself, or if one vector system is equivalent to a certain second system and the latter is, in turn, equivalent to a third system, the third system is also equivalent to the first system.

Ergo. If two finite vector systems $a_1 \ldots e_m$ and $b_1 \ldots b_n$ are linearly independent and equivalent, then $m = n$.

Definition 3.10. If vectors' transformation E leaves them unchanged, then this transformation is called *identical* and the matrix E of this transformation is the *identity matrix* :

$$E = \begin{Vmatrix} 1 & 0 & 0 & \ldots \\ 0 & 1 & 0 & \ldots \\ 0 & 0 & 1 & \ldots \\ \ldots & \ldots & \ldots & \ldots \end{Vmatrix}$$

Definition 3.11. The transformation that transforms an image A_x (where x is the vector obtained as a result of the transformation of A), back to the vector x is called the *inverse transformation* and is denoted A^{-1}.

Ergo. The inverse transformation can be expressed by the identity matrix: $AA^{-1} = E$.

Definition 3.12. The transformation in linear space, which in a normal basis is expressed by orthogonal matrices, is the *orthogonal transformation* ; if the matrices are symmetric, we speak of *symmetric transformations* .

Definition 3.13. The transformation A is orthogonal if and only it does not change the modules (lengths) of the vectors, i.e. for any vector x,

$$(Ax, Ax) = (x, x) \text{ or } (Ae_i, Ae_i) = (e_i, e_i)$$

Accordingly, for orthogonal vectors, $(Ae_i, Ae_i) = 0$, where $i \neq j$.

Definition 3.14. Whatever the symmetric transformation of a plane is, there exist two mutually perpendicular eigenvectors for it that belong to the same or different eigenvalues.

Definition 3.15. Symmetric transitions are always reduced to the tension or compression of the plane in two mutually perpendicular directions.

Definition 3.16. A rank of a set M of vectors is the maximum number of linearly independent variables.

As mathematical tools for calculations in linear vector spaces, the methods of higher algebra are used [51], in whose description we use the techniques proposed in the lectures of E. Fermi and R. Feynman, who were not only outstanding scientists, but also brilliant teachers [48, 50].

Rule 3.1. Only identical matrices may be added and subtracted:

$$\|a_{ik}\| \pm \|b_{ik}\| = \|a_{ij} \pm b_{ik}\| = \|c_{ik}\|$$

$$\begin{Vmatrix} a_{11} & a_{12} & a_{13} \\ a_{21} & a_{22} & a_{23} \end{Vmatrix} \pm \begin{Vmatrix} b_{11} & b_{12} & b_{13} \\ b_{21} & b_{22} & b_{23} \end{Vmatrix} = \begin{Vmatrix} a_{11} \pm b_{11} & a_{12} \pm b_{12} & a_{13} \pm b_{13} \\ a_{21} \pm b_{21} & a_{22} \pm b_{22} & a_{23} \pm b_{23} \end{Vmatrix}$$

$$= \begin{Vmatrix} c_{11} & c_{12} & c_{13} \\ c_{21} & c_{22} & c_{23} \end{Vmatrix}$$

Rule 3.2. Matrices multiplication is carried out by the rule 'row × column', i.e. $c_{ik} = \sum\limits_j a_{ij} b_{ik}$:

$$\begin{Vmatrix} a_{11} & a_{12} & a_{13} \\ a_{21} & a_{22} & a_{23} \\ a_{31} & a_{32} & a_{33} \end{Vmatrix} \times \begin{Vmatrix} b_{11} & b_{12} & b_{13} \\ b_{21} & b_{22} & b_{23} \\ b_{31} & b_{32} & b_{33} \end{Vmatrix} = \begin{Vmatrix} c_{11} & c_{12} & c_{13} \\ c_{21} & c_{22} & c_{23} \\ c_{31} & c_{32} & c_{33} \end{Vmatrix}$$

where, e.g. c_{23} is a product of the second raw and the third column, i.e.

$$c_{23} = a_{21} \cdot b_{13} + a_{22} \cdot b_{23} + a_{23} \cdot b_{33}$$

Corollary 3.1. *from Rule 3.2*: The result of multiplication of different matrices can be visually represented [50]:

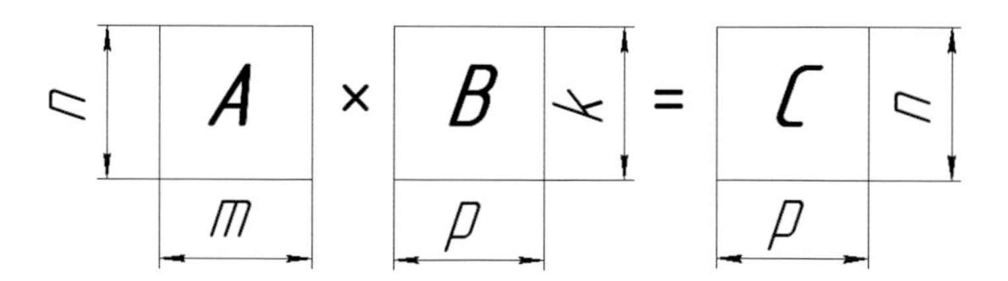

3.2 Linear Spaces in Quantum Mechanics

Despite the highest degree of mathematization of theoretical physics, there is a difference in physical and mathematical forms of description. In mathematics, the consideration of any matter may start with the introduction of a specific set of axioms, such as 'parallel lines never intersect', or vice versa, 'parallel lines can cross', and based on that, rigorous logical consequences are obtained. A physician is deprived of this freedom. He cannot 'assign' axioms for his theory because their corollaries must certainly satisfy the experimentally observed characteristics in the theoretical model of the object. A physicist cannot know in advance what kind of maths is required for that description. In physical theories, the inductive component is less important than deductive and it can never be carried out on the level of rigor that characterizes mathematical models. As a consequence of this, we will follow the path of successive refinement of calculating models and the concepts used in them, not initially detailing the properties of the vector space of a quantum system; only formally applying the rules of 3.1 and what we already know about vectors that describe the states of quantum objects.

In accordance with those properties, as well as Def. 3.5, ket-vectors may be written as a linear combination:

$$\begin{gathered} \left|R\right\rangle = c_1\left|A_1\right\rangle + c_2\left|A_2\right\rangle + \dots \\ c_1\left|A\right\rangle + c_2\left|A\right\rangle + \dots = (c_1 + c_2 + \dots)\left|A\right\rangle \end{gathered} \tag{3.1}$$

We introduce the linear scalar functions for ket-vectors $\phi(\left|A\right\rangle)$, which should have the following properties:

$$\begin{gathered} \phi\left(c_1\left|A\right\rangle + c_2\left|B\right\rangle\right) = c_1\phi\left(\left|A\right\rangle\right) + c_2\phi\left(\left|B\right\rangle\right) \\ (\phi_1 + \phi_2)\left(\left|A\right\rangle\right) = \phi_1\left(\left|A\right\rangle\right) + \phi_2\left(\left|A\right\rangle\right) \\ (c\phi)\left(\left|A\right\rangle\right) = c\phi\left(\left|A\right\rangle\right) \text{ for any vector } \left|A\right\rangle \end{gathered} \tag{3.2}$$

Then the parameters of ϕ will also form a linear vector space that is dual (conjugate) to the initial one. Under Dirac's terminology, the vectors of this space are called 'bra-vectors' $\langle B|$:

$$\begin{gathered}\phi\left(|A\rangle\right)=\langle B|A\rangle\\ \langle B|\left\{|A_1\rangle+|A_2\rangle\right\}=\langle B|A_1\rangle+\langle B|A_2\rangle\\ \left\{\langle B_1|+\langle B_2|\right\}|A\rangle=\langle B_1|A\rangle+\langle B_2|A\rangle\\ \left\{\langle B|C\rangle\right\}|A\rangle=C\langle B|A\rangle\end{gathered}\tag{3.3}$$

If $\langle B|A\rangle=0$ for any ket-vector $|A\rangle$, then $\langle B|=0$.

(3.1)–(3.3) set the rules of working with quantum vectors used in Section 2.4. Scalar products of two bra-vectors and two ket-vectors are not possible. Instead, we assume that between the spaces of the vectors (ket) and the co-vectors (bra) an antilinear bijection of type $\langle|=F\left(|\rangle\right)$ is set, for which

$$\begin{gathered}F\left(|A\rangle\right)=F\left(|A\rangle\right)+F\left(|B\rangle\right)\\ F\left(\langle C|A\rangle\right)=CF\left(|A\rangle\right)\end{gathered}\tag{3.4}$$

Now each state A may be correlated not only to the vector $|A\rangle$, but also its co-vector $\langle A|$. Therefore, to describe the two states A and B, it is logical to use the appropriate vectors and co-vectors that can form scalar products:

$\langle B|A\rangle$ – linear in $|A\rangle$ and antilinear in $|B\rangle$

$\langle A|B\rangle$ – antilinear in $|A\rangle$ and linear in $|B\rangle$, and, under (2.24), $\langle A|B\rangle^*$ is linear in $|A\rangle$ and antilinear in $|B\rangle$.

As a part of what is written, we should make the following notes:

Note 3.1. For the same result, we may not refer to 'bypass configuration' with conjugate space, but directly introduce the Hermitian metric with positive determinacy into the linear vector space of ket-vectors.

Note 3.2. When comparing the findings of quantum mechanics with the experimental data, the states would act as two 'manifestations' in terms of their observation and in term of their 'preparing' (Dirac's term). These two sides of the physical manifestation of the state best meet the dual mathematical model of bra- and ket-vectors related by (3.4). It should be emphasized that a physical description using vectors must be perfectly symmetric, which can be interpreted as the isotropy of time, which is expressed in the requirement of a Hermitian character of a quantum space metric. As for the condition of positive metric of the physical systems space, it is necessary to save the probabilistic interpretation of quantum theory.

The scalar product of conjugate vectors allows to introduce a so-called 'norm' in both spaces assuming that

$$\left\| \left| A \right\rangle \right\| = \left\| \left\langle A \right| \right\| = +\sqrt{\left\langle A \middle| A \right\rangle} \tag{3.5}$$

With the concept of norm, the procedure of correlating the vectors to the states of the quantum system may be specified, i.e. specific states are associated by only normalized vectors for which $\left\| \left| A \right\rangle \right\| = 1$.

The next object that is logical to define for the vector space of the physical system is a linear vector-function $\left| G \right\rangle$, which is regarded as an effect of the linear operator $\hat{\alpha}$ on the argument $\left| A \right\rangle$:

$$\begin{gathered} \left| G \right\rangle = \alpha \left| A \right\rangle \\ \hat{\alpha}\left\{ c_1 \left| A_1 \right\rangle + c_2 \left| A_2 \right\rangle + \ldots \right\} = c_1 \hat{\alpha} \left| A_1 \right\rangle + c_2 \hat{\alpha} \left| A_2 \right\rangle + \ldots \\ \text{If } \hat{\alpha} = 0, \text{ then } \hat{\alpha} \left| A \right\rangle = 0 \text{ for any } \left| A \right\rangle \\ \left\{ \alpha_1 + \alpha_2 \right\} \left| A \right\rangle = \alpha_1 \left| A \right\rangle + \alpha_2 \left| A \right\rangle \\ \left\{ \alpha\beta \right\} \left| A \right\rangle = \alpha \left\{ \beta \left| A \right\rangle \right\} \end{gathered} \tag{3.6}$$

To determine the effect of the operator on the co-vector, consider the scalar product $\left\langle B \middle| \hat{\alpha} \middle| A \right\rangle$, which is a linear scalar function of vector $\left| A \right\rangle$, and is thus equal to the scalar product of $\left| A \right\rangle$ and co-vector $\left\langle B' \right|$:

$$\left\langle B' \middle| A \right\rangle = \left\langle B \middle| \hat{\alpha} \middle| A \right\rangle \tag{3.7}$$

Co-vector $\left\langle B' \right|$ is a linear function of co-vector $\left\langle B \right|$:

$$\left\langle B' \right| = \left\langle B \right| \hat{\alpha} \tag{3.8}$$

Comparing (3.7) and (3.8), it is natural to propose that $\hat{\alpha} = \hat{\alpha'}$, then

$$\left(\left\langle B \middle| \hat{\alpha} \middle| A \right\rangle \right) = \left\langle B \right| \left(\hat{\alpha} \middle| A \right\rangle \Big) \text{ for any } \left| A \right\rangle \tag{3.9}$$

Under (3.9), the associative rule is carried out in between the vectors in our vector space and the product (3.9) may be rewritten as (2.20), giving it a corresponding physical meaning: $\left\langle B \middle| \bar{\alpha} \middle| A \right\rangle$ and called it the matrix element of the operator $\hat{\alpha}$.

We introduce a concept of conjugate operator. For that, we associate the vector $\left| Q \right\rangle$ with the co-vector $\left\langle P \right| \alpha = \left\langle Q \right|$, which is antilinear to the bra-vector $\left\langle P \right|$ and thus linear to the ket-vector $\left| P \right\rangle$. Then, under Rule (2.24),

$$\begin{gathered} \alpha^{+} \left| P \right\rangle = \left| Q \right\rangle, \text{ as } \left\langle Q \right| = \left\langle P \right| \alpha \\ \left\langle B \middle| \alpha^{+} \middle| P \right\rangle = \left\langle P \middle| \alpha \middle| B \right\rangle \\ \alpha = \alpha^{+}, \ (\alpha^{+})^{+} = \alpha \end{gathered} \tag{3.10}$$

The special form of linear operators is obtained if we form the products of ket- and bra-vectors $\left| A \right\rangle \left\langle B \right|$:

$$\begin{gathered} \left\{ \left| A \right\rangle \left\langle B \right| \right\} \left| P \right\rangle = \left| A \right\rangle \left\{ \left\langle B \middle| P \right\rangle \right\} \\ \left\langle Q \right| \left\{ \left| A \right\rangle \left\langle B \right| \right\} = \left\{ \left\langle Q \middle| A \right\rangle \right\} \left\langle B \right| \\ \left\{ \left| A \right\rangle \left\langle B \right| \right\}^{+} = \left| B \right\rangle \left\langle A \right| \end{gathered} \tag{3.11}$$

Such operators have been called projectors P:

$$\begin{gathered} \hat{P}^{+} = \hat{P} \\ \hat{P}^{2} = \hat{P} \\ \left\langle A \middle| P \middle| A \right\rangle = \left\langle A \middle| P \cdot P \middle| A \right\rangle = \left\| P \middle| A \right\rangle \Big\|^{2} \geq 0 \end{gathered} \tag{3.12}$$

We introduced the operators in the state vector space following the natural course of a mathematical, not a physical, thought. Indeed, it is hard to find any sense in correlation of one state to another or to find a room to the operation of transformation of the old system state into the new one. Among few similarities that we have, one may represent the changes by symmetry transformation as a 'set of elements of a certain nature, for which some group operation is defined'. Nonetheless, we make a number of assumptions based on previously formulated general physical principles. For example, the dynamic variables in classical theory correspond to the linear operators of the state vector space in quantum mechanics. And the real dynamic variables correspond to the Hermitian operators. The next assumption suggests that the dynamic variables, affecting the state of the system, convert it to a different state. Evidently, it is possible to find common grounds for the dynamic variables introduced this way, in the sense of classical physics.

Continuing to build the linear vector space of a quantum system as defined in (3.12), we introduce an idea of perpendicularity, which is a generalization of the concept associated with our usual space. We say that the bra- and ket-vectors are orthogonal if their scalar products are equal to zero. Two bra-vectors or two ket-vectors are orthogonal if the scalar product of one of them and the conjugate vector of another is equal to zero. The two states of a quantum system are orthogonal if the vectors that characterize these states are orthogonal to one another.

The length of the bra-vector $\left\langle A \right|$ or its conjugate ket-vector $\left| A \right\rangle$ is defined as $+\sqrt{\left\langle A \middle| A \right\rangle}$. If a system state is given and it is necessary to introduce the bra- or ket-vectors corresponding to this state, only the vector direction is defined; the vector itself is defined only to an arbitrary factor. It can be conveniently chosen so that the vector length is equal to one. This procedure is called normalization in quantum mechanics. The state vector is not fully defined even in this case, since it is not prohibited to multiply it by a coefficient equal module one, i.e. by any number exp $(i\gamma)$ with the real exponent γ without changing the vector length. This coefficient is called a phase factor.

Definition 3.17. The space of bra- and ket-vectors that have a finite length and a finite scalar product is called a *Hilbert space* [52].

Definition 3.18. *Hilbert space* is a complex vector space, which is a complete infinite-dimensional Euclidean space. This means the existence of a set of elements in a Hilbert space, for which, in addition to the traditional operations for the vector space, a complex function of a pair of arguments is defined that satisfy a set of axioms to define the commutative, associative and distributive properties of the variables that make up the complex-valued vector function arguments.

We would like to finish this part of the chapter with a quote from the famous Dirac's monograph that clearly reflects the essence of the problem of this section: "...we set up an algebraic scheme involving certain abstract quantities of three kinds, namely bra-vectors, ket-vectors, and linear operators, and we expressed some of the fundamental laws of quantum mechanics in terms of them. It would be possible to continue to develop the theory in terms of these abstract quantities and to use them for applications to particular problems. However, for some purposes it is more convenient to replace the abstract quantities by sets of numbers with analogous mathematical properties and to work in terms of these sets of numbers. The procedure is similar to using coordinates in geometry and has the advantage of giving one greater mathematical power for solving of particular problems" [44].

All the above is achieved by switching to a matrix representation of quantum quantities. We state without proof a number of theorems that illustrate the features of the matrix representation of quantum mechanical equations [49].

Theorem 3.1. *In a finite manifold, the most common linear operator is reduced to a homogenous linear combination:*

$$A_\kappa = \sum_{i=1}^{n} a_{\kappa i} f_i \tag{3.13}$$

Definition 3.19. A square $(n \times n)$ matrix is called *Hermitian* (or self-conjugate) if each of its elements is conjugate to the element that is symmetric with respect to the main diagonal of the matrix; in other words, the matrix A is Hermitian if $a_{j\kappa} = a^*_{\kappa i}$, the diagonals of Hermitian matrices are either zeros or real numbers.

Theorem 3.2. *Raising an Hermitian matrix to any power gives an Hermitian matrix.*

Theorem 3.3. *The determinant of an Hermitian matrix is a real number.*

Theorem 3.4. *If A is an Hermitian matrix, then its reverse matrix A^{-1} is also Hermitian.*

Theorem 3.5. *Let $F(x)$ be a real function of a real variable x, such that it can be compared to a matrix $F(A)$, i.e. define a function of the matrix A, and if A is Hermitian then the function $F(A)$ is also an Hermitian matrix.*

As part of the matrix representation of quantum mechanics, consider a problem of the eigenvalues of the operators composed with Hermitian matrices. Let there be a Hermitian operator $\hat{A} = \hat{A}^{+}$ whose eigenvalues problem $\hat{A}\psi = a\psi$ is rewritten in matrix form:

$$\begin{gathered} a_{11}\psi_1 + a_{22}\psi_2 + \cdots + a_{1n}\psi_n = a\psi_1 \\ a_{21}\psi_1 + a_{22}\psi_2 + \cdots + a_{2n}\psi_n = a\psi_2 \\ \cdots\cdots\cdots\cdots\cdots\cdots\cdots\cdots\cdots \\ a_{n1}\psi_1 + a_{n2}\psi_2 + \cdots + a_{nn}\psi_n = a\psi_n \end{gathered} \tag{3.14}$$

The system (3.14) is called a *system of secular equations* , the coefficients $a_{i\kappa}$ are the elements of an Hermitian matrix A. The system has a solution if the determinant of the system (3.14) is

$$\begin{vmatrix} a_{11} - a & a_{12} & \dots & a_{1n} \\ a_{21} & a_{22} - a & \dots & a_{2n} \\ \dots & \dots & \dots & \dots \\ a_{n1} & a_{n2} & \dots & a_{nn} - a \end{vmatrix} = 0 \tag{3.15}$$

The Hermitian matrix operator has n real eigenvalues, some of which may coincide. The eigenvalues $a_1, a_2, \ldots, a_n$, which are the roots of the Eq. (3.15), correspond to the wave eigenfunctions $\psi_1, \psi_2, \ldots, \psi_n$.

Theorem 3.6. *If all n roots of a secular equation are multiples of one, each eigenvalue corresponds to a unique wave eigenfunction ψ.*

Rule 3.3. To construct the wave function ψ, substitute the eigenvalue a_s instead of the value a into the secular determinant (3.15). Then n algebraic adjuncts of any of the raw determinant will be proportional to the component of the vector ψ_s [51].

Rule 3.4. The matrix operator is completely defined by assigning eigenvectors and corresponds to the eigenvalues.

Definition 3.20. A matrix Q is called *unitary* if it has the following property: $Q^{+}Q = 1$ or $Q^{+} = Q^{-1}$.

Theorem 3.7. *If a matrix T is unitary, the following relation is valid:* $\left\langle Tf \middle| Tg \right\rangle = \left\langle f \middle| g \right\rangle$.

Theorem 3.8. *If a matrix T is unitary and ψ_s is an orthogonal system of n vectors, the result of transformation of $T\psi_s$ will also be an orthogonal vectors system ϕ_s: $T\psi_s = \phi_s$.*

Theorem 3.9. *The coordinates transformation is described by the same matrix as the inverse transformation of the basis vectors.*

Definition 3.21. A *spur* or a *trace* of a square matrix is equal to the sum of its diagonal components.

Theorem 3.10. *The trace of the operator is equal to the sum of its eigenvalues.*

Theorem 3.11. *If a degenerated matrix B commutes with a diagonal matrix A, the matrix B should also be diagonal, i.e. if*

$$A = \begin{Vmatrix} a_1 & 0 & 0 & \dots & \dots \\ 0 & a_2 & 0 & \dots & \dots \\ 0 & 0 & a_3 & 0 & \dots \\ \dots & \dots & \dots & \dots & \dots \\ 0 & 0 & 0 & \dots & a_n \end{Vmatrix}, \textit{ then } B = \begin{Vmatrix} b_{11} & b_{12} & 0 & 0 & \dots \\ b_{21} & b_{22} & 0 & 0 & \dots \\ 0 & 0 & b_{33} & b_{34} & \dots \\ 0 & 0 & b_{43} & b_{44} & \dots \\ \dots & \dots & \dots & \dots & \dots \end{Vmatrix} \tag{3.16}$$

Supplement. (under Fermi [50]): Let A and B be Hermitian matrices and let them commute with one another: $[AB] = 0$; if there exist the matrices $A' = T^{+}AT$ and $B' = T^{+}BT$ and they also commute, then, provided the matrix A is not degenerate, the matrix B is diagonal. If A is degenerate, B' has the form (3.16) and its secular equation decomposes into simpler equation whose degree is a multiple of the degeneracy of the eigenvalues of the matrix A.

Definition 3.22. *Transposition* is a swapping of the columns and the rows in a matrix:

$$f_i = \begin{Vmatrix} f_1 & f_2 & f_3 & \dots \end{Vmatrix} \to \bar{f}_i = \begin{Vmatrix} f_1 \\ f_2 \\ \dots \end{Vmatrix} \tag{3.17}$$

Theorem 3.12. *Hermitian conjugation of the matrices is done by transposition and complex conjugation simultaneously.*

In a number of problems that quantum physics faces, it is very convenient to convert the operators in quantum-mechanical equations into the matrix form. At the same time, the orthonormality must be observed and as basis vectors, the wave eigenfunctions of an often used operator are chosen, e.g. the Hamiltonian. As an example, consider a one-dimensional quantum system with the generalized coordinate $q = x$. The system of orthonormal basis wave function in the matrix form is as follows:

$$T = \begin{Vmatrix} \psi_1(x') & \psi_2(x') & \dots \\ \psi_1(x'') & \psi_2(x'') & \dots \\ \dots & \dots & \dots \end{Vmatrix} \tag{3.18}$$

The matrix T obtained in (3.18) is doubly infinite; its vector function $f(x)$ is

$$f(x) = \sum_n \phi_n \psi_n(x) \tag{3.19}$$

where the coefficient ϕ_n is the scalar product $\left\langle \psi_n \middle| f'(x) \right\rangle$; $f'(x)$ are the old components of the vector function $f(x)$; ϕ_n are its new components.

As an example, convert into the explicit matrix form the most often used operators [49]:

a) the matrices of the Hamiltonian $\hat{H}$ after unary transformations:

$$\hat{H} = \begin{Vmatrix} 1/2\hbar\omega & 0 & 0 & 0 & 0 & \dots \\ 0 & 3/2\hbar\omega & 0 & 0 & 0 & \dots \\ 0 & 0 & 5/2\hbar\omega & 0 & 0 & \dots \\ 0 & 0 & 0 & 7/2\hbar\omega & 0 & \dots \\ \dots & \dots & \dots & \dots & \dots & \dots \end{Vmatrix} \tag{3.20}$$

b) the matrices of the coordinate and momentum operators:

$$\begin{gathered} q = \sqrt{\frac{\hbar}{2m\omega}} \begin{Vmatrix} 0 & \sqrt{1} & 0 & 0 & \dots \\ \sqrt{1} & 0 & \sqrt{2} & 0 & \dots \\ 0 & \sqrt{2} & 0 & \sqrt{3} & \dots \\ 0 & 0 & \sqrt{3} & 0 & \dots \\ \dots & \dots & \dots & \dots & \dots \end{Vmatrix} \\ p = \sqrt{\frac{\hbar m\omega}{2}} \begin{Vmatrix} 0 & -i\sqrt{1} & 0 & 0 & \dots \\ i\sqrt{1} & 0 & -i\sqrt{2} & 0 & \dots \\ 0 & i\sqrt{2} & 0 & -i\sqrt{3} & \dots \\ \dots & \dots & \dots & \dots & \dots \\ \dots & \dots & \dots & \dots & \dots \end{Vmatrix} \end{gathered} \tag{3.21}$$

where $\hbar$ is the Planck's constant; ω is the frequency; m is the mass; i is the imaginary unit $\sqrt{-1}$.

c) the matrix of the momentum operator M:

$$M_x = \hbar \begin{Vmatrix} l & 0 & 0 & 0 & \dots \\ 0 & l-1 & 0 & 0 & \dots \\ \dots & \dots & \dots & \dots & \dots \\ 0 & 0 & 0 & \dots & -l \end{Vmatrix}$$

$$M_y = \frac{\hbar}{2} \begin{Vmatrix} 0 & ib_l & 0 & \dots & \dots \\ ib_l & 0 & ib_{l-1} & \dots & \dots \\ \dots & \dots & \dots & \dots & \dots \\ 0 & 0 & 0 & \dots & -ib_{-l+1} \\ 0 & 0 & 0 & -ib_{-l-1} & 0 \end{Vmatrix} \quad (3.22)$$

$$M_z = \frac{\hbar}{2} \begin{Vmatrix} 0 & b_l & 0 & 0 & 0 \\ b_l & 0 & b_{l-1} & 0 & 0 \\ 0 & b_{l-1} & 0 & b_{l-1} & 0 \\ \dots & \dots & \dots & \dots & \dots \\ 0 & 0 & 0 & 0 & b_{-l+1} \end{Vmatrix}$$

where $b_s = \sqrt{(l+s)(l+1-s)}$; l is the orbital quantum number; s is the spin quantum number.

3.3 Relation between the Quantum Values for Dimensional Subspaces

We begin our description of physics of quantum systems as is. From the above, we may give the following definition of a 'quantum system':

Definition 3.23. A *quantum system* generally refers to any physical system described in terms of states with the use of concepts, such as wave function, state vector, density matrix. This approach is the most complete description of physical processes. Classical science is only a special case of quantum theory [53].

Further development of the axioms of quantum mechanics requires the extension of the definition of the 'state' concept with the use of space-time concepts. Based on Def. 2.21 that belongs to P. Dirac and is based on the relationship of the concepts of 'state' and 'undisturbed motion', which can be regarded as a sign of closure of the physical system. Therefore, it is possible to define the state as a characteristic that best describes a closed physical system. Keeping in mind that we are talking about quantum states, it is necessary to supplement the above by a clarification that this description is in the chosen basis, which is formed by state vectors in a Hilbert space. The Hilbert space of states where our quantum system is defined characterizes the set of states of the latter. It is given by a set of eigenstates with their vectors when we are interested in a particular case (Def. 3.18, Theorem 3.1, etc.).

Thus, depending on the task, a basis is chosen and various state vectors for the same system are written. For example, the problem requires a study on the dynamics of changes in the coordinates of a quantum particle. Then infinite-dimensional Hilbert space is chosen because the coordinates are continuous quantities and each point of the space corresponds to a separate state of the quantum system. The completeness of the quantum description is not to simultaneously describe all that is possible for the system studied; the completeness of the description in the sense of quantum mechanics is the ability to have a complete description of the object of interest within a certain set of its states. Having noted the state vector in different bases (2.23), we analyze the studied system from different angles, considering its numerous manifestations.

The analysis of the 'quantum aspects' of the measurement procedure, the detection of 'quantum non-locality' effects described in Section 2.2 allow description of the physical system in terms of information. In this case, the physical processes of strengthening and weakening the quantum entanglement between the components (subsystems) of the quantum object are regarded as an information exchange between the system itself and its environment. In this sense, the background information on the quantum system is a single information field, which contains the data on all possible implementations of the internal structures of the system. S. Doronin compared the quantum information system with an undeveloped photographic paper that contains a specific set of potential pictures [40]. In quantum theory, this corresponds to the location of a closed physical system in the non-local state because of the inability (as defined in its 'closed' status) to carry out the procedure of reduction of the wave vector.

In our consideration, the information is a quantitative characteristic of the physical system. It includes information on the system that we get in the process of measuring its physical parameters (Def. 2.18).

It is not a question of any characteristic, but of specific quantities that have an unambiguous definition, to which also the information that can take a 'different look' may be attributed. According to a number of known theoretical physicists, advocated by S. Doronin, "...the essense of quantum information and, at the same time, its exclusive feature is that it is absolutely consistent with the requirement of the primary substance... Any physical system can be thought of as consisting of two qualitatively different components. One piece is an eternal, indestructible core that exists beyond space-time and the other is a dense body, the shell, through which the system implements a particular state of its core" [40].

In this specification, each energetic structure of a quantum system can be regarded as a subspace, the continuum that has its own event space depending, for example, on the average energy density of that structure. In other words, each component of a quantum system is in its own event space and in various degrees of entanglement in accordance with its physical characteristics. Since the subsystems are able to interact, there arises an extremely important problem of the relationship between the various parameters that characterize them as physical objects. Since the metric of these individual 'partial' continua are regulated by energy creating the subsystem space, Hamiltonians play a special role in constraint equations.

The basis for finding such relations is the statement known as the Wigner theorem:

Theorem 3.13. (Wigner theorem) . *If the operator $\hat{A}$ commutes with all the bases transformations in a group, then, accordingly, the matrix elements $\left\langle \phi_i \middle| \hat{A} \middle| \phi_\kappa \right\rangle$, choosing the basis operators, must be equal to zero if the amplitudes ϕ_i and ϕ_κ correspond to irreducible representations [31].*

$$\left\langle \phi_i \middle| \hat{A} \middle| \phi_\kappa \right\rangle = a_{i\kappa}\delta_{i\kappa}$$
$$\delta_{i\kappa} = \begin{cases} 1 & i = \kappa \\ 0 & i \neq \kappa \end{cases} \tag{3.23}$$

where $a_{i\kappa}$ is a matrix consisting of real numbers.

Another important provision to find the relations that connect the states of various subspaces of a quantum system is a corollary from Lemma 2.1. According to Lemma 2.1, if a physical system is degenerate, the essential requirement is its simultaneous coexistence with measurable physical parameters. This means that various results of the measurements carried out on a quantum system cannot occur simultaneously because of inequality of their commutators to zero. Therefore, the impact of an act of measurement on the quantum system is able to simultaneously 'materialize' just one of the informational sets of its possible manifestations (say, Everett worlds). So, between the similar parameters of the quantum quantities related to various spaces, non-local connections manifested in one of the most important laws of quantum mechanics, the Heisenberg uncertainty principle [37], which, in notation of Lemma 2.1, can be written as

$$\left[\hat{L}\hat{M}\right] = -i\hbar \tag{3.24}$$

where $\hat{L}, \hat{M}$ are operators of physical quantities of diverse dimensions; $\hbar$ is the Planck's constant.

(3.23) and (3.24) complement and extend each other, linking the eigenvalues of different representations of a quantum system. Further, theoretical constructs are impossible without returning to the quantum multidimensional world of the Hilbert space.

Let us introduce the direct product of the ket-vectors called the tensor product [49]. From each subspace of the quantum system, we take only one vector $\left|A_1\right\rangle\left|A_2\right\rangle\ldots\left|A_N\right\rangle$ where N is the number of subspaces (or the number of degrees of freedom), such that

$$\begin{aligned}\left|A_1\right\rangle\ldots\left\{c_1\left|A_1\right\rangle + c_2\left|B_1\right\rangle\right\}\ldots\left|A_N\right\rangle = c_1\left|A_1\right\rangle\ldots\left|A_N\right\rangle \\ + c_2\left|A_1\right\rangle\ldots\left|B_i\right\rangle\ldots\left|A_N\right\rangle\end{aligned} \tag{3.25}$$

for any $1 \leq i \leq N$.

Then the tensor products may be considered, such as

$$\left|A_1\right\rangle\ldots\left|A_N\right\rangle = \left|A_1\ldots A_N\right\rangle \tag{3.26}$$

as an element of a new vector space formed as a 'linear shell' of the products (3.25)–(3.26), which is called the direct product $\left\{\left|A_1\right\rangle\otimes\cdots\otimes\left|A_N\right\rangle\right\}$ of the subspaces $\left\{\left|A_1\right\rangle\right\}\cdot\left\{\left|A_2\right\rangle\right\}\ldots\left\{\left|A_N\right\rangle\right\}$ [49].

The scalar product in the 'big' outer space $\left\{\left|A_1\right\rangle\right\} \otimes \cdots \otimes \left\{\left|A_N\right\rangle\right\}$ is first defined for the pair of vectors and co-vectors as (3.26):

$$\left\langle B_1 \ldots B_N \middle| A_1 \ldots A_N \right\rangle = \left\langle B_1 \middle| A_1 \right\rangle \ldots \left\langle B_N \middle| A_N \right\rangle \tag{3.27}$$

Having introduced the corresponding state vectors, we start to 'build' the operators for the outer space and initially use a quantity denoted by Dirac as ξ_i. It is an operator that acts in i^{th} small space. We assume these operators unambiguously match the operators Ξ_i from the 'big' space:

$$\Xi_i \left| A_1 \ldots A_N \right\rangle = \left| A_1 \right\rangle \ldots \xi_i \left| A_i \right\rangle \ldots \left| A_N \right\rangle \tag{3.28}$$

Formula (3.28) needs to be commented:

Note 3.3. In terms of physics, the operators ξ_i and Ξ_i are identical, since they are the square representatives of the same physical quantity. In terms of mathematics, they differ because they exist in different vector spaces.

The emergence of new components of spatial structures with bounded metric requires the introduction of new definitions.

Definition 3.24. A finite-dimensional linear vector space is called *configuration space* .

Definition 3.25. For a single material point, the configuration space coincides with the usual Euclidean space or the Minkowski continuum; in all other cases it is not true. If a physical system consists of n points, the configuration space that determines it consists of $3n$ dimensions.

Based on the relationship between the quantum parameters and the features of 'geometry' of the space where these parameters are defined, we derive the uncertainty principle, very important for further analysis. In Section 2.1, the Heisenberg formula was derived from the idea of the indivisibility of the instrument – quantum object system. Here, upon logical constructions based on the above material, consider a quantum system in a configuration space described by the ket-vector $\left|A\right\rangle$ normalized by one (3.5): $\left\| \left|A\right\rangle \right\|^2 = 1$. In accordance with the conclusions of Section 2.4, the measurements give only the average values that, upon (2.16)–(2.19), are represented as

$$\begin{aligned}\bar{q} &= \left\langle A \middle| q \middle| A \right\rangle \\ \bar{p} &= \left\langle A \middle| p \middle| A \right\rangle\end{aligned} \tag{3.29}$$

where $\bar{q}$ and $\bar{p}$ are the average values (observable) of the physical parameters q and p.

The quantization conditions that we have do not impose any additional restrictions on the average values, so nothing forbids us to introduce an assumption that $\bar{q} = 0$ and $\bar{p} = 0$. The features of real physical measurements determine a certain scatter of the data obtained characterized by the value of the standard deviation:

$$\begin{gathered}(\Delta q)^2 = \overline{(q - \bar{q})}^2 = \bar{q}^2 - 2q\bar{q} + \bar{q}^2 = \bar{q}^2 \\ \text{Since } \bar{q} = 0, \text{ therefore, } \bar{q}^2 = 0, \text{ so} \\ (\Delta p)^2 = \bar{p}^2\end{gathered} \tag{3.30}$$

Let us introduce the state vector $\left| B \right\rangle$ using a procedure, the physical meaning of which will be regarded later as:

$$\left| B \right\rangle = \frac{p - i\alpha q}{\sqrt{2\pi\hbar}} \left| A \right\rangle = a \left| A \right\rangle \tag{3.31}$$

Consider the vector from (3.31):

$$\begin{aligned}0 \leq \left\| \left| B \right\rangle \right\|^2 &= \left\langle A \middle| \frac{p + i\alpha q}{\sqrt{2\pi\hbar}} \cdot \frac{p - i\alpha q}{\sqrt{2\pi\hbar}} \middle| A \right\rangle = \frac{1}{2\pi h} \left\{ \left\langle A \middle| p^2 \middle| A \right\rangle \right. \\ &\quad \left. - i\alpha \left\langle A \middle| pq - qp \middle| A \right\rangle + \alpha^2 \left\langle A \middle| q^2 \middle| A \right\rangle \right\}\end{aligned} \tag{3.32}$$

The condition of non-negativity of the vector $\left| B \right\rangle$ (3.32) is met if $\alpha^2(\Delta q)^2 - \hbar\alpha + (\Delta p)^2 \geq 0$, at which the discriminant of this inequality should be less than or equal to zero:

$$\begin{gathered}\hbar^2 - 4(\Delta q)^2(\Delta p)^2 \leq 0 \\ \text{or} \\ \Delta q \Delta p \geq \frac{\hbar}{2}\end{gathered} \tag{3.33}$$

The (3.33) is the canonical notation form of the Heisenberg uncertainty principle. Consider the equality conditions of (3.33) $\hbar/2$. This is possible only if the norm $||B||^2$ is equal to zero, i.e. if $\Big|A\Big\rangle = 0$ and $\Big|B\Big\rangle = 0$. This state is called *physical vacuum* [53]. Then the coordinate and momentum representations of a vacuum are of the form [44]:

$$\begin{aligned} \Big\langle q'\Big|0\Big\rangle = c_0(q') = \sqrt{\frac{\alpha}{\pi\hbar}}\exp\left(-\frac{\alpha q'^2}{2\hbar}\right) \\ \Big\langle p'\Big|0\Big\rangle = d_0(p') = \frac{1}{\sqrt{\pi\alpha\hbar}}\exp\left(-\frac{p'^2}{2\alpha\hbar}\right) \end{aligned} \tag{3.34}$$

where α is an arbitrary real parameter introduced before into (3.31).

So, a whole collection of states characterized by the lowest degree of uncertainty, called zero fluctuations, are linked with the state of physical vacuum:

$$\begin{aligned} \Delta q = \sqrt{\frac{\hbar}{2\alpha}} \\ \Delta p = \sqrt{\frac{\hbar\alpha}{2}} \end{aligned} \tag{3.35}$$

We found that the states described by the vector $\Big|A\Big\rangle$ contain the following properties: $\Big\langle A\Big|A\Big\rangle = 1$; $\bar{q}_A = \Big\langle A\Big|q\Big|A\Big\rangle = 0$; $\bar{p}_A = \Big\langle A\Big|p\Big|A\Big\rangle = 0$; $\Delta p\Delta q = \frac{\hbar}{2}$; $\Delta p = \sqrt{\overline{\Delta p}^2}$; $\Delta q = \sqrt{\overline{\Delta q}^2}$; $\overline{\Delta p^2} = \Big\langle A\Big|(p-\bar{p})^2\Big|A\Big\rangle$; $\overline{\Delta q^2} = \Big\langle A\Big|(q-\bar{q}^2)\Big|A\Big\rangle$ contains the essence of the physical vacuum state that we will further denote $\Big|A\Big\rangle = \Big|0\Big\rangle$.

The existence of zero fluctuations reveals the 'mechanism' of the uncertainty principle. For example, the fact that the particles move in a real material environment does not stop even with the absolute zero temperature. It is often said that at absolute zero, only the thermal Brownian motion stops. However, to preserve the fundamental law of the microuniverse – the uncertainty principle – these particles have to participate in a motion that delocalize their position in space. This motion is called zero fluctuation. The 'spatial' analysis of the Heisenberg principle 'explains' the origin of that type of motion as a change in the depths of an 'enigmatic' environment actually studied as a form of existence of matter – physical vacuum [53]. The corollary from the Wigner theorem implies that the relation between the states of a quantum system (between its different spatial 'incarnations') is only possible in the form of non-local correlations characterizing the quantum entanglement of those subsystems. This behavior determines the interaction 'mechanism' of different 'quantum worlds'.

The states with the least uncertainty are most close to their classical analog, since the impact of the Heisenberg principle (3.33) is minimized. Such states, if the equalities $q = \bar{q}$ and $p = \bar{p}$ are met, are called coherent. In other words, coherent states are states of least uncertainty in the values of the coordinate and momentum at the given mean values. In terms of S. Doronin, "Coherent states are a superposition of pure states, i.e. overlapping of individual states that a closed system can be in. Coherence is a consistency of behavior of individual components of a physical system by means of non-local correlations" [40].

Coherent states are associated with the processes taking place 'inside' the physical vacuum. They are the essence of the eigenstates of the operator called an annihilation operator:

$$a\left|\bar{p},\bar{q}\right\rangle = \bar{a}\left|p,q\right\rangle \tag{3.36}$$

In accordance with the previously established requirements to the description of quantum states, we think that the eigenvectors (3.36) of the coherent states we found should form a normalized (but not necessarily orthonormal) system. It can be shown that

$$\begin{aligned}\left\langle\bar{p}'\bar{q}'\middle|\bar{p}\bar{q}\right\rangle &= e^{-\frac{\bar{a}\bar{a}^{*}+\bar{a}^{*}\bar{a}}{2}}\left\langle 0\middle|e^{\bar{a}^{*}a}\cdot e^{\bar{a}a^{*}}\middle|0\right\rangle \\ &= e^{-\frac{1}{2\pi}\left[\frac{(p'-\bar{p})^{2}}{2\alpha}+\frac{\alpha(q'-\bar{q})^{2}}{2}\right]}e^{\frac{i}{2\pi}(p'q'-q'p')}\end{aligned} \tag{3.37}$$

Expanding the 'quantum space parameters' into the eigenvectors of the annihilation operator a, we get:

$$\left\langle n\middle|\bar{p}',\bar{q}'\right\rangle = e^{-\frac{\bar{a}^{*}\bar{a}}{2}}\left\langle n\middle|e^{\bar{a}\bar{a}^{*}}\middle|0\right\rangle = \frac{1}{\sqrt{n!}}\bar{a}^{n}e^{-\frac{a^{*}a}{2}} \tag{3.38}$$

We will use the equations in (3.38) form later. Here we note the following: a system of coherent states with fixed parameters α and arbitrary values of $\bar{q}$ and $\bar{p}$ is superabundant. Therefore, to construct it, it is enough to use some countable sequence of coherent states. The Hilbert space, as defined in Def. 3.24, can be considered a configuration phase space, which significantly facilitates further construction of the quantum model of a tribosystem.

3.4 Spatial Transformations in Quantum Mechanics

The data presented in Section 3.3 allow to simplify several ideas of the spatial organization of a quantum physical system. In particular, the vector space containing the given quantum object can be represented as a set of all its possible states. If they are infinite, we have a Hilbert space; if their number is finite, the quantum system is defined in the configuration space. The basis of the wave function corresponding to Def. 3.3 determines the metric of the space. Properly quantum system is a multidimensional information field, which is converted into its 'material-substance' forms only in conditions of decoherence, i.e. breaking the quantum non-locality as a result of the interaction of the quantum system with the environment (the instrument). As a result of decoherence, the multidimensional quantum system is 'projected' onto our real classic world, while losing a certain number of dimensions.

Since the state vectors determine the metric of the configuration space, the operation of 'rewriting' this vector from one basis to another in the form of (2.23) should characterize the interspatial transition made by the quantum system as a result of either decoherence or, in the reverse process, recoherence [40].

Quantum mechanics 'presents' accurate formulae that allow writing the state vectors in a definite basis [49]:

$$\begin{aligned}
\left\langle n \middle| q \right\rangle &= \frac{(-i)^n}{\sqrt{2^n n!}} H_n\left(\sqrt{\frac{2\alpha}{\pi}}q\right)\left(\frac{\alpha}{\pi\hbar}\right)^{\frac{1}{4}} e^{-\frac{\alpha}{2\pi}q^2} \\
\left\langle n \middle| p \right\rangle &= \frac{1}{\sqrt{2^n n!}} H_n\left(\frac{1}{\sqrt{\hbar\alpha}}p\right)\frac{1}{\left(\frac{\pi\hbar}{\alpha}\right)^{1/4}} e^{-\frac{p^2}{2\pi\hbar}} \\
\left\langle q' \middle| q \right\rangle &= \delta(q' - q) \\
\left\langle p' \middle| p \right\rangle &= \delta(p' - p)
\end{aligned} \tag{3.39}$$

where H_n is the Hermite polynomial [50]; δ is the Dirac function.

$$H_n(\xi) = (-1)^n e^{\xi^2} \frac{d^n}{d\xi^n} e^{-\xi^2} = \begin{cases} H_0(\xi) = 1 \\ H_1(\xi) = 2\xi \\ H_2(\xi) = 4\xi^2 - 2 \end{cases}$$

$$\frac{dH_n(\xi)}{d\xi} = 2nH_{n-1}(\xi)$$

$$\int\limits_{-\infty}^{\infty} H_n^2(\xi) e^{-\xi^2} d\xi = \sqrt{\pi} 2^n n! \tag{3.40}$$

$$\int\limits_{-\infty}^{\infty} H_n(\xi) e^{-\xi^2} e^{ip\xi} d\xi = i^n \sqrt{\pi} e^{-\frac{p^3}{4}}$$

$$H_n(\xi) = 2\xi H_{n-1}(\xi) - 2(n-1) H_{n-2}(\xi)$$

$$\delta(x) \frac{1}{2\pi} \int\limits_{-\infty}^{\infty} e^{i\kappa x} dx$$

$$\delta(x) = \lim_{a \to \infty} \frac{\sin \alpha x}{\pi x}$$

$$\int\limits_{-\infty}^{\infty} f(x) \delta(x-a) dx = f(a) \tag{3.41}$$

$$-\int\limits_{-\infty}^{\infty} f(x) \delta'(x-a) dx = f'(a)$$

The basis of the basic (big) multidimensional space can be represented as a set of one-dimensional representations (3.39):

$$\left\langle n_1 \ldots n_N \middle| q_1 \ldots q_N \right\rangle = \left\langle n_1 \middle| q_1 \right\rangle \ldots \left\langle n_N \middle| q_N \right\rangle$$

$$= \frac{(-i)^{\sum n_i}}{\sqrt{2^{\sum n_i} n_1! \ldots n_N!}} \left(\frac{\alpha}{\pi\hbar}\right)^{\frac{N}{4}} \cdot \exp\left[-\frac{\alpha}{2\hbar}\sum_i q_i^2\right] H_{n_1}\left(\sqrt{\frac{\alpha}{\hbar}} q_1\right)$$

$$\ldots H_{n_H}\left(\sqrt{\frac{\alpha}{\hbar}} q_N\right)$$

$$\left\langle n_1 \ldots n_N \middle| p_1 \ldots p_N \right\rangle = \left\langle n_1 \middle| p_1 \right\rangle \ldots \left\langle n_N \middle| p_N \right\rangle$$

$$= \frac{1}{\sqrt{2^{\sum n_i} n_1! \ldots n_N!}} \frac{1}{(\frac{\pi\hbar}{\alpha})^{\frac{N}{4}}} \exp\left[-\frac{1}{2\pi\hbar}\sum_i p_i^2\right] H_{n_1}\left(\frac{1}{\sqrt{\hbar\alpha}} p_1\right)$$

$$\ldots H_{n_N}\left(\frac{1}{\sqrt{\hbar\alpha}} p_N\right) \tag{3.42}$$

Relations (3.39) and (3.42) are written accurately within the phase factor $e^{i\phi}$ [50].

The increase or decrease in spatial dimensions is done by a consistent effect on the state vector by the birth operator and the annihilation operator, which we used in (3.31) in the derivation of the uncertainty principle [50]. The operators are introduced as

$$a = \frac{p - i\alpha q}{\sqrt{2\alpha\hbar}},\ a^+ = \frac{p + i\alpha q}{\sqrt{2\alpha\hbar}} \tag{3.43}$$

Accordingly, from (3.43) one may see the relation between the value of the coordinates and momentum operators and the operators of birth and annihilation:

$$q = \sqrt{\frac{\hbar}{2\alpha}} \frac{a^+ - a}{i},\ p = \sqrt{\frac{\alpha\hbar}{2}}\left(a^+ + a\right) \tag{3.44}$$

The birth and annihilation operators have the following properties:

$$\lfloor aa^+ \rfloor = aa^+ - a^+a = 1$$

Accordingly,

$$
\begin{gathered}
aa^{+}a\Big|n\Big\rangle = a\cdot n\Big|n\Big\rangle,\ a^{+}a^{+}a\Big|n\Big\rangle = a^{+}n\Big|n\Big\rangle \\
\Big(a^{+}a+1\Big)a\Big|n\Big\rangle = na\Big|n\Big\rangle,\ \Big(a^{+}a-1\Big)a^{+}\Big|n\Big\rangle = na^{+}\Big|n\Big\rangle \\
a^{+}a\Big\{a\Big|n\Big\rangle\Big\} = \Big(n-1\Big)\Big\{a\Big|n\Big\rangle\Big\} \\
a^{+}a\Big\{a^{+}\Big|n\Big\rangle\Big\} = \Big(n+1\Big)\Big\{a^{+}\Big|n\Big\rangle\Big\} \\
a\Big|n\Big\rangle = \Big|n-1\Big\rangle \\
a^{+}\Big|n\Big\rangle = \Big|n+1\Big\rangle \\
a^{+}a\Big|n\Big\rangle = n\Big|n\Big\rangle
\end{gathered}
\tag{3.45}
$$

The vector production $a^{+}a$ is called particles number operator. The chosen names of these operators very accurately disclose their physical essence, which is shown in (3.45). The action of the operator a decreases the number of bases, reducing the dimension (number of particles, etc.) by one unit. On the contrary, the effect of its conjugate operator a^{+} increases this number by one particle. This process is described by $\Big\langle n'\Big|a\Big|n''\Big\rangle$ as $\Big\langle n'\Big|a^{+}\Big|n''\Big\rangle$, which, according to Feynman's interpretation, shows how the birth and annihilation operators act on space (the system) described by the basis $\Big|n''\Big\rangle$ transforming it into $\Big|n'\Big\rangle$. This action can be written in explicit matrix form [49]:

$$
\begin{gathered}
\Big\langle n'\Big|a\Big|n''\Big\rangle = \begin{Vmatrix} 0 & \sqrt{1} & 0 & 0 & 0 & \dots \\ 0 & 0 & \sqrt{2} & 0 & 0 & \dots \\ 0 & 0 & 0 & \sqrt{3} & 0 & \dots \\ 0 & 0 & 0 & 0 & \sqrt{4} & \dots \\ \dots & \dots & \dots & \dots & \dots & \dots \end{Vmatrix} \\
\Big\langle n'\Big|a^{+}\Big|n''\Big\rangle = \begin{Vmatrix} 0 & 0 & 0 & 0 & \dots \\ \sqrt{1} & 0 & 0 & 0 & \dots \\ 0 & \sqrt{2} & 0 & 0 & \dots \\ 0 & 0 & \sqrt{3} & 0 & \dots \\ 0 & 0 & 0 & \sqrt{4} & \dots \\ \dots & \dots & \dots & \dots & \dots \end{Vmatrix}
\end{gathered}
\tag{3.46}
$$

In the matrices (3.46) the components $a_{11} = 0$ corresponding to the state of the physical vacuum (3.34), (3.35) deserve special attention. This state is obtained by the effect of the annihilation operator on a system with a single basis or a single particle. In relation to the state of the physical vacuum, the following can be written:

$$\begin{aligned} a\left|0\right\rangle &= 0\left|0\right\rangle \\ \left\|\left|0\right\rangle\right\| &= 1 \end{aligned} \tag{3.47}$$

The ket-vector $\left|0\right\rangle$ that characterizes the vacuum is an eigenvector of a paired operator a^+a called particle number operator with zero eigenvalue, as all the other vectors differ from it by an integer. Therefore, the eigenvectors of the particle number operator are integers; $\left|n\right\rangle$ is a set of vectors $n = 0, 1, 2, 3\ldots$ form the basis in the vector space of quantum system states with a single degree of freedom. Then any operator $\hat{A}$ has a 'representative' in the configuration space as a matrix $\left\langle n'\middle|A\middle|n''\right\rangle$ by which the operator is written as a double infinite sum of basis vectors and co-vectors [49].

$$A = \sum_{n'}\sum_{n''}\left\langle n'\middle|A\middle|n''\right\rangle\left|n'\right\rangle\left\langle n''\right| \tag{3.48}$$

One may ascertain that a set of dyadic products of the form $\left|n'\right\rangle\left\langle n''\right|$ acts as a basis in the space of linear operators.

The state of the physical vacuum is nondegenerate. All the eigenvectors of the particle number operator a^+a are obtained from the vacuum by consistent action of the birth operator a^+ on it:

$$\left|n\right\rangle = c_n\left(a^+\right)\left|0\right\rangle \tag{3.49}$$

where $c_n(a^+)$ is a normalized factor in vector decomposition that arose for the first time in (3.34); $c_n(a^+)$ is normalized to one.

Then it can be written:

$$\begin{aligned} \left\|\left|n\right\rangle\right\|^2 &= c_n^* c_n n! \\ c_n &= \frac{1}{\sqrt{n!}}, \\ \left|n\right\rangle &= \frac{1}{\sqrt{n!}}\left(a^+\right)^n\left|0\right\rangle \end{aligned} \tag{3.50}$$

Let us give a definition to a concept 'representation by a representative'.

Definition 3.26. Any complete set of commuting observables determines a complete orthogonal family in a vector space $\left|A\right\rangle$. Quantum mechanics says about the choice of a definite basis as of choice of representative. For an arbitrary state vector denoted by $\left|A\right\rangle$, one may write, using the relations (3.11) and (3.12), which determine the projector, the following:

$$\left|A\right\rangle = \sum_{\xi_i} \dots d\xi_n p_{\xi_1 \dots \xi_n}\left|A\right\rangle = \sum_{\xi_i} \dots d\xi_n \left|\xi_1 \dots \xi_n\right\rangle\left\langle\xi_1 \dots \xi_n\right|A\right\rangle \tag{3.51}$$

The formula (3.51) is a 'sum on complete system' and a set of the number in (3.51) completely determines the vector $\left|A\right\rangle$. This set is called a 'vector representative in the given representation'. Accordingly, the representative of the co-vector (bra) $\left\langle A\right|$ is a set of scalar products with basis vectors:

$$\left\langle A\middle|\xi_1 \dots \xi_n\right\rangle = \left\langle\xi_1 \dots \xi_n\middle|A\right\rangle^* \tag{3.52}$$

By definition, the representative of the vector $\left|A\right\rangle$ and its conjugated co-vector, written in one representation, are complex conjugate to one another [44].

In accordance with (3.48) and Def. 3.26, we write the birth and annihilation operators with projectors [49]:

$$\begin{aligned} a^+ &= \sum_{n'} \sqrt{n'+1}\left|n'+1\right\rangle\left\langle n'\right| \\ a &= \sum_{n''} \sqrt{n''+1}\left|n''\right\rangle\left\langle n''+1\right| \end{aligned} \tag{3.53}$$

Similarly, we write the degrees of these operators:

$$\left(a^+\right)^2 = \sum_{n'} \sqrt{(n'+2)(n'+1)}\left|n'+2\right\rangle\left\langle n'\right|$$

$$\left(a\right)^2 = \sum_{n''} \sqrt{(n''+2)(n''+1)}\left|n''\right\rangle\left\langle n''+2\right|$$

$$\left(a\right)^{\nu_1} = \sum_{n'} \sqrt{\frac{(n'+\nu_1)!}{n'!}}\left|n'+\nu_1\right\rangle\left\langle n'\right|$$

$$\left(a\right)^{\nu_2} = \sum_{n''} \sqrt{\frac{(n''+\nu_2)}{n''!}}\left|n''\right\rangle\left\langle n''+\nu_2\right|$$

$$\left(a^+\right)^{\nu_1} a^{\nu_2} = \sum_{n'}\sum_{n''} \sqrt{\frac{(n'+\nu_1)!(n''+\nu_2)!}{n'!n''!}}\left|n'+\nu_1\right\rangle\left\langle n'\middle|n''\right\rangle\left\langle n''+\nu_2\right| \tag{3.54}$$

Let us introduce a concept of the projector into the concept of vacuum. By definition of this operator,

$$p_0 = \left|0\right\rangle\left\langle 0\right| = \sum_{\nu} \frac{(-1)^{\nu}}{\nu!}\left(a^+\right)^{\nu} a^{\nu} = e^{-a^+a} \tag{3.55}$$

The projector representative, by (3.48) is

$$\begin{aligned}\left\langle n'\middle|p_0\middle|n''\right\rangle &= \delta_{n'_0}\delta_{n''_0} = \sum_{\nu_1}\sum_{\nu_2} c_{\nu_1\nu_2}\left\langle n'\middle|\left(a^+\right)^{\nu_1} a^{\nu_2}\middle|n''\right\rangle \\ &= \sum_{\nu_1}\sum_{\nu_2} \delta_{(n'-\nu_1)(n''-\nu_2)}\sqrt{\frac{n'!n''!}{(n'-\nu_1)!}}c^0_{\nu_1\nu_2}\end{aligned} \tag{3.56}$$

where $c^0_{\nu_1\nu_2} = \dfrac{(-1)^{\nu_1}}{\nu_1!}\delta_{\nu_1\nu_2}$; $\delta_{n_0} = \sum\limits_{i=0}^{n} \dfrac{n!}{(n-\nu_1)!}\dfrac{(-1)^{\nu_1}}{\nu_1!}$.

Combining (3.55) and (3.56), we can obtain the formula for an arbitrary projector p_n.

$$p_n = \left|n\right\rangle\left\langle n\right| = \frac{1}{n!}\left(a^+\right)^n\left|0\right\rangle\left\langle 0\right|a^n = \frac{1}{n!}\left(a^+\right)^n p_0 a^n = \frac{\left(a^+a\right)^n}{n!}e^{-a^+a} \tag{3.57}$$

The relations (3.49), (3.53)–(3.57) have a very interesting physical interpretation we dealt with earlier in [40, 53]. It is associated with a not changing its classic analog possibility to 'materialize particles out of the void'. It has two approaches, one of which may be regarded as a traditional for classical mechanics looks at the birth of particles 'out of nothing', based on Dirac's works on relativistic quantum theory and his concept of states of physical vacuum [44, 53]. The second approach is relatively new and controversial as it requires the introduction of a large number of additional concepts.

According to Dirac, the physical vacuum is not an absolute void but an environment, a new state of matter built of particles in a virtual state. Virtual particles have negative energy and are separated from ordinary particles with a band gap and, therefore, according to Dirac, 'virtually unobservable'. Actually the physical vacuum is presented in the form of negative energy levels completely filled with virtual particles, as well as completely free levels of 'usual' positive energy, separated by energy gap [53] (Fig. 3.1). Imposing a portion of energy exceeding this gap to a virtual particle results in the fact that is passes to normal levels, seems to appear in our world, 'materializes out of nothing'.

P. Dirac

The development of these ideas allowed creating an entirely new theory of interaction, in which the forces between particles arise because of an exchange of virtual quanta [28]. Having materialized in our world, a virtual particle leaves a trace in the physical vacuum in the form of an antiparticle, which is regarded by one of the founders of the new force theory R. Feynman as the particle that moves 'backwards in time' [54]. The existence of physical vacuum as an objective reality has numerous experimental confirmations and may therefore be considered a reality.

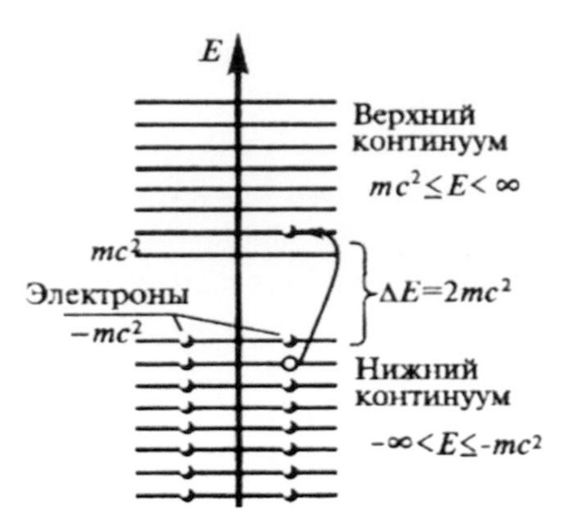

Figure 3.1: Dirac's vacuum model. The arrow shows the electron-positron pair birth process

The second approach is radically different. In its terms, the atomic-molecular structure of bodies can no longer be regarded as the fundamental basis for building our world. Particles as primary, self-sufficient elements of reality are secondary manifestations of the decoherence of non-local quantum states (Fig. 3.2). The reverse process of transition of a physical object from local to non-local form – the recoherence – is also possible [40]. Therefore, in accordance with these views, quantum theory implies the existence of a deeper 'not manifested, not local' reality, of which our world in only one of the projections. In fact, all our 'real-field' universe is immersed in a reality of a higher level [40].

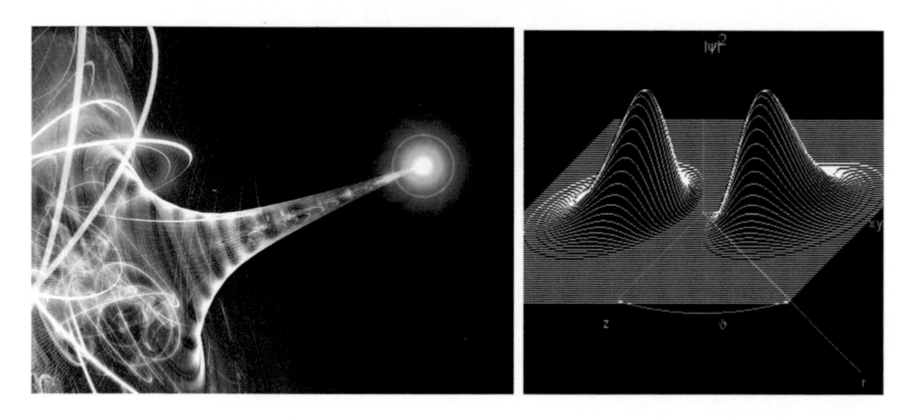

Figure 3.2: Decoherence – the birth out of 'nothing' (quantum pictures)

The introduced concept of non-local quantum reality has no classical analog and cannot be described by the methods of classical physics.

The non-local world is a void quantitatively measured in terms of quantum information. This void is not a vacuum, not an empty space occupying a certain volume; it does not even have this volume nor spacetime. This is a single information field and the processes of decoherence (materialization) are local manifestations of this information as a result of an act of quantum measurement.

Any physical system within this approach, as mentioned earlier, has a structure consisting of two parts, two qualitatively different components: an eternal, indestructible core that exists beyond time and space and a shell, by which a physical system implements one of its possible states hidden in the core.

As said an 'adept' at the new approach to micromechanisms of quantum forces, S. Doronin, "Decoherence is a manifestation of hidden information... in a certain classical reality, the projection of the developed picture... All these projections remain within the universe and are not visible from the outside. There is no single general scene for all. Different parts of the system take part as 'decorations' in their performances, and all the spectators are distributed in their 'interests', parameters and energy characteristics as local objects-participants" [38]. The advantages of this approach are universality and flexibility, which unite all the possible theories of being, from de Broglie to Everett, and can in this sense be considered models of the 'theory of everything' [27]. The problem is that this universality is achieved by introducing the concept of 'non-local reality', which is attributed to mythical characteristics that exceed, for example, ether or caloric. The science experience shows that the explanation of Nature by introducing too fantastic notions that do not have even close analogs in the surrounding reality and contrary to all known scientific doctrines are later discarded and replaced with 'usual' material 'details' of the objective reality. Nonetheless, the 'non-local source of quantum reality' has its supporters, and it is likely to be good because nowadays there is simply no comprehensive theory (except the Bible) that would tell us about the beginning of time. By the way, the Bible also 'confesses the information view' on the creation of the world, '...in the beginning was the Word'.

3.5 Transitions between Subspaces in a Quantum Tribosystem

The material presented in the previous sections gives a first definition associated with the main objective of this book—creation of a general tribosystem model that would satisfy all known empirical data. This definition should characterize properly tribosystem within the problem and the available information.

Definition 3.27. *Properly tribosystem* is a multidimensional quantum object in a non-local highly degenerate state; the physical and technical characteristics of the tribosystem manifest only the process of interaction (act of quantum measurement) with a physical instrument, a quantum tribometer, as a consequence of the reduction of quantum characteristics of the tribosystem itself.

Definition 3.28. *Manifested tribosystem* is the result of the decoherence of tribosystem states realized in the form of fixed physical, chemical and mechanical parameters.

Definition 3.29. *Quantum tribometer* is a set of experimental methods of influence on properly tribosystem resulting in the appearance of the manifested tribosystem.

As a result of the decoherence caused by the interaction of properly tribosystem (hereinafter tribosystem) with a quantum tribometer, the 'materialization' of the manifested tribosystem occurs. This materialization (or 'localization' in terms of theory of non-local quantum effects) can be regarded as a 'projection' of non-local quantum information substance of the tribosystem from the multidimensional Hilbert space onto the Minkowski four-dimensional continuum. This transformation of linear spaces may be represented as a system of linear equations corresponding to Def. 3.5. The number of equations in the system is equal to the number of dimensions of the quantum system and the number of linear combination terms of each equation is determined by the basis of the space where our tribosystem materializes. For the Minkowski continuum, there are four terms: three space coordinates (x, y, z) and one time coordinate t.

$$
\begin{aligned}
&\left|A_1\right\rangle = a_{11}\left|q_1\right\rangle + a_{12}\left|q_2\right\rangle + a_{13}\left|q_3\right\rangle + a_{14}\left|q_4\right\rangle \\
&\left|A_2\right\rangle = a_{21}\left|q_1\right\rangle + a_{22}\left|q_2\right\rangle + a_{23}\left|q_3\right\rangle + a_{24}\left|q_4\right\rangle \\
&\cdots\cdots\cdots\cdots\cdots\cdots\cdots\cdots\cdots\cdots\cdots\cdots \\
&\left|A_n\right\rangle = a_{n1}\left|q_1\right\rangle + a_{n2}\left|q_2\right\rangle + a_{n3}\left|q_3\right\rangle + a_{n4}\left|q_4\right\rangle
\end{aligned}
\tag{3.58}
$$

The states described in the system are orthonormal. Therefore, the off-diagonal coefficients a_{ij} whose indices differ by more than one should be set to zero. By Defs. 3.17 and 3.19, the matrix that describes the state of properly tribosystem should be a square, four rows by four columns, table. Therefore, it is expected that the linear space where the tribosystem is defined in a non-local state has sixteen dimensions (4×4). Researches of superexcited states of the friction surfaces substance discovered a shell structure of triboplasma that has 15–16 more or less stable quantum levels [37]. Since energy is an important, if not decisive, factor in the arrangement of space-time metric, it is expectable that the quantum structure of the triboplasma somehow reflects that metric (Fig. 3.3).

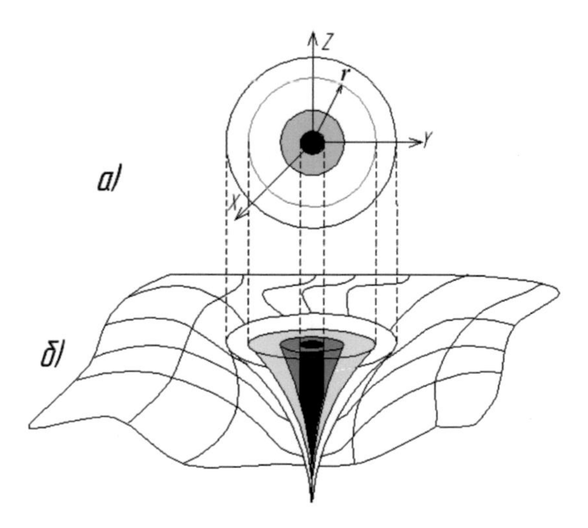

Figure 3.3: Relation of the space-time continuum distortion with triboplasma generation: (a) is a metric representation of the triboplasma shell structure (the shading intensity indicates its density, the darker the denser); (b) is the distortion of the space-time continuum relevant to the energy dense

Anyway, the probability that the number of quantum shells matches the number of dimensions of the tribosystem configuration space is most likely non-random.

We get the matrix A_{ij} of the system (3.58):

$$A_{ij} = \begin{Vmatrix} a_{11} & a_{12} & 0 & 0 \\ a_{21} & a_{22} & a_{23} & 0 \\ 0 & a_{32} & a_{33} & a_{34} \\ 0 & 0 & a_{43} & a_{44} \end{Vmatrix} \tag{3.59}$$

This matrix resembles matrix B from (3.16). Therefore, in accordance with Theorem 3.11, the matrix A_{ij} that describes the tribosystem defines that the quantum tribosystem in a non-local state is a degenerate quantum object, which confirms Def. 3.27.

If the structure of the levels describing the linear spaces of quantum objects 'localization' corresponds, to some extent, to 'visible' images of those spaces, the 'chess structure' of the matrix (3.59) is possible:

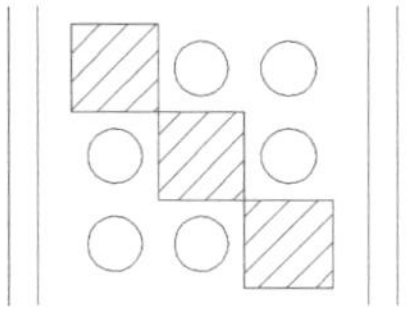

(3.60)

It represents the potentiality of the tribosystem to choose its manifested image that shows the 'true' friction image by reducing the number of variations of independent dimensions.

The Wigner theorem 3.13 and Lemma 2.1 imply that if the states of the initial and the final systems commute, a condition is met

$$\left[\hat{H}\hat{A}\right] = 0 \tag{3.61}$$

where $\hat{H}$ is the Hamiltonian of the initial system; $\hat{A}$ is any operator describing the final form of the system.

Assume that the Hamiltonian $\hat{H}$ refers to properly tribosystem in a 16-dimensional configuration space. In terms of the notation used in Eq. (3.28), we denote the Hamiltonian of the system subjected to decoherence into the Minkowski space after the first measurement by h_1 and after the second by h_2. From (3.61) we note:

$$\begin{gathered}[Hh_1] = [Hh_2] = Hh_1 - h_1 H = Hh_2 - h_2 H \\ Hh_1 + h_2 H = Hh_2 + h_1 H\end{gathered} \tag{3.62}$$

which in the matrix form looks $\left\langle n \middle| Hh_1 + h_2 H \middle| n \right\rangle =$ $\left\langle n \middle| Hh_2 + h_1 H \middle| n \right\rangle$

In accordance with (3.3), we rewrite the last equation in (3.62) as follows:

$$\begin{gathered}\left\langle n \middle| Hh_1 + h_2 H \middle| n \right\rangle = \left\langle n \middle| Hh_1 \middle| n \right\rangle + \left\langle n \middle| h_2 H \middle| n \right\rangle \\ \left\langle n \middle| Hh_2 + h_1 H \middle| n \right\rangle = \left\langle n \middle| Hh_2 \middle| n \right\rangle + \left\langle n \middle| h_1 H \middle| n \right\rangle\end{gathered} \tag{3.63}$$

Substituting (3.63) in (3.62), we get

$$\left\langle n \middle| Hh_1 \middle| n \right\rangle + \left\langle n \middle| h_2 H \middle| n \right\rangle = \left\langle n \middle| Hh_2 \middle| n \right\rangle + \left\langle n \middle| h_1 H \middle| n \right\rangle \tag{3.64}$$

By Def. 2.4 and 2.5 and their corollaries,

$$\begin{gathered}\left\langle n \middle| Hh \middle| n \right\rangle = \left\langle n \middle| H \middle| n \right\rangle \left\langle n \middle| h \middle| n \right\rangle \\ \left\langle n \middle| hH \middle| n \right\rangle = \left\langle n \middle| h \middle| n \right\rangle \left\langle n \middle| H \middle| n \right\rangle\end{gathered} \tag{3.65}$$

(3.64), taking into account the transformations in (3.65) is rewritten as

$$\begin{gathered}\left\langle n \middle| H \middle| n \right\rangle \left\langle n \middle| h_1 \middle| n \right\rangle + \left\langle n \middle| h_2 \middle| n \right\rangle \left\langle n \middle| H \middle| n \right\rangle \\ = \left\langle n \middle| H \middle| n \right\rangle \left\langle n \middle| h_2 \middle| n \right\rangle + \left\langle n \middle| h_1 \middle| n \right\rangle \left\langle n \middle| H \middle| n \right\rangle \\ \text{or} \\ \left\langle n \middle| H \middle| n \right\rangle \left\langle n \middle| h_1 \middle| n \right\rangle - \left\langle n \middle| H \middle| n \right\rangle \left\langle n \middle| h_2 \middle| n \right\rangle \\ = \left\langle n \middle| h_1 \middle| n \right\rangle \left\langle n \middle| H \middle| n \right\rangle - \left\langle n \middle| h_2 \middle| n \right\rangle \left\langle n \middle| H \middle| n \right\rangle\end{gathered} \tag{3.66}$$

From Dirac's notation we pass to the common mathematical equation form:

$$H_{nn} h_1^{mm} - H_{nn} h_2^{mm} = h_1^{mm} H_{nn} - h_2^{mm} H_{nn} \tag{3.67}$$

The matrix of the Hamiltonian H_{nn} describes the configuration space that has sixteen dimensions; the matrices h^{mm} are associated with four-dimensional Minkowski space, so explicitly (3.67) should look

$$\begin{Vmatrix} H_{11} & H_{12} & H_{13} & H_{14} \\ H_{21} & H_{22} & H_{23} & H_{24} \\ H_{31} & H_{32} & H_{33} & H_{34} \\ H_{41} & H_{42} & H_{43} & H_{44} \end{Vmatrix} \times \begin{Vmatrix} h_{11} & h_{12} \\ h_{21} & h_{22} \end{Vmatrix} = \begin{Vmatrix} c_{11} & c_{12} \\ c_{21} & c_{22} \\ c_{31} & c_{32} \\ c_{41} & c_{42} \end{Vmatrix}$$
$$\begin{Vmatrix} h_{11} & h_{12} \\ h_{21} & h_{22} \end{Vmatrix} \times \begin{Vmatrix} H_{11} & H_{12} & H_{13} & H_{14} \\ H_{21} & H_{22} & H_{23} & H_{24} \\ H_{31} & H_{32} & H_{33} & H_{34} \\ H_{41} & H_{42} & H_{43} & H_{44} \end{Vmatrix} = \begin{Vmatrix} c'_{11} & c'_{12} & c'_{13} & c'_{14} \\ c'_{21} & c'_{22} & c'_{23} & c'_{24} \end{Vmatrix} \tag{3.68}$$

(in matrix transformations, Rule 3.2 and its corollaries were used).

The coefficients c_{ij}, c'_{ji} of the output matrices (3.68) are found by Rule 3.2. Substituting the matrices of form (3.68) into (3.67), we get its explicit form.

Right-hand side of (3.67):

$$\begin{Vmatrix} c_{11} & c_{12} \\ c_{21} & c_{22} \\ c_{31} & c_{32} \\ c_{41} & c_{42} \end{Vmatrix} h_1 - \begin{Vmatrix} c'_{11} & c'_{12} \\ c'_{21} & c'_{22} \\ c'_{31} & c'_{32} \\ c'_{41} & c'_{42} \end{Vmatrix} h_2 = \begin{Vmatrix} c_{11} - c'_{11} & c_{12} - c'_{12} \\ c_{21} - c'_{21} & c_{22} - c'_{22} \\ c_{31} - c'_{31} & c_{32} - c'_{32} \\ c_{41} - c'_{41} & c_{42} - c'_{42} \end{Vmatrix}$$

Left-hand side of (3.67):

$$\begin{Vmatrix} c'_{11} & c'_{12} & c'_{13} & c'_{14} \\ c'_{21} & c'_{22} & c'_{23} & c'_{24} \end{Vmatrix} h_1 - \begin{Vmatrix} c''_{11} & c''_{12} & c''_{13} & c''_{14} \\ c''_{21} & c''_{22} & c''_{23} & c''_{24} \end{Vmatrix} h_2$$
$$= \begin{Vmatrix} c'_{11} - c''_{11} & c'_{12} - c''_{12} & c'_{13} - c''_{13} & c'_{14} - c''_{14} \\ c'_{21} - c''_{21} & c'_{22} - c''_{22} & c'_{23} - c''_{23} & c'_{24} - c''_{24} \end{Vmatrix}$$

The indices h_1 and h_2 at the matrix brackets mean that the matrix coefficients were calculated with the Hamiltonian h_1 in one case and with h_2 in another.

By substituting the obtained results in (3.67), we find its explicit form:

$$\begin{Vmatrix} c_{11} - c'_{11} & c_{12} - c'_{12} \\ c_{21} - c'_{21} & c_{22} - c'_{22} \\ c_{31} - c'_{31} & c_{32} - c'_{32} \\ c_{41} - c'_{41} & c_{42} - c'_{42} \end{Vmatrix} = \begin{Vmatrix} c'_{11} - c''_{11} & c'_{12} - c''_{12} & c'_{13} - c''_{13} & c'_{14} - c''_{14} \\ c'_{21} - c''_{21} & c'_{22} - c''_{22} & c'_{23} - c''_{23} & c'_{24} - c''_{24} \end{Vmatrix} \tag{3.69}$$

Equation (3.69) shows that there was a transposition of the matrices and change in modules of the vectors that characterize the observables. Under Def. 3.21 and Theorem 3.12, this change in matrix structure can be interpreted as an evidence that the quantum states that satisfy the right-hand and left-hand sides of (3.69) will be conjugate (though not necessarily Hermitian conjugate). In geometric interpretation, it may look like that consistent changes result in a rotation of the coordinate axes by 90 (Fig. 3.4). Therefore, both subspaces that arise as a result of the interaction of a tribosystem and a quantum tribometer are invisible to one another. If any relation between them is possible, it could only be the correlations associated with the entangled states of the quantum system. Their form is subject to the relations of the form (3.24).

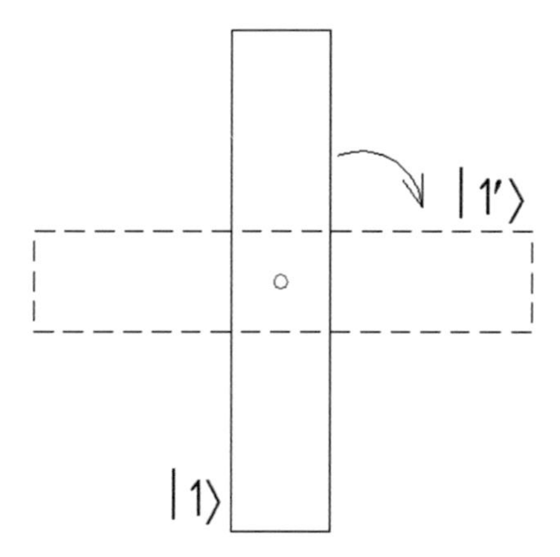

Figure 3.4: Symbolic representation of 'coexistence' of two 'subspaces' in a tribosystem formed by consistent acts of measurement

If a quantum system described by a state vector $\left|\psi\right\rangle$ makes a turn by an angle ϕ, then, in terms of theory of unitary transformations [49], the initial vector $\left|\psi\right\rangle$ is transformed into another ket-vector $\left|\tilde{\psi}\right\rangle$:

$$\begin{aligned}\left|\tilde{\psi}\right\rangle &= \exp\left[-\frac{i}{\hbar}\phi(nM)\right]\left|\psi\right\rangle = \exp\left[-\frac{i}{\hbar}\cdot\frac{\hbar}{2}(n\sigma)\right]\left|\psi\right\rangle \\ &= \exp\left[-\frac{i}{2}(n\sigma)\right]\left|\psi\right\rangle = \left\{\cos\left[\frac{\phi}{2}(n\sigma)\right] - i\sin\left[\frac{\phi}{2}(n\sigma)\right]\right\}\end{aligned} \tag{3.70}$$

where M is the torque operator; σ is the vector spin operator.

Since we are interested in observable quantities, i.e. in only what is registered by direct measurements, only the real part of (3.70) is chosen:

$$\mathrm{Re}\left|\tilde{\psi}\right\rangle = \cos\left[\frac{\phi}{2}(n\sigma)\right]\left|\psi\right\rangle \tag{3.71}$$

In our case, $(n\sigma)^2 = n^2 = 1$, i.e. $n\sigma = 1$, then (3.71) is simplified as

$$\begin{gathered}\mathrm{Re}\left|\tilde{\psi}\right\rangle = \cos\frac{\phi}{2}\left|\psi\right\rangle \\ \text{and since } \phi = \pi/2 \\ \mathrm{Re}\left|\tilde{\psi}\right\rangle = \left|\tilde{\psi}\right\rangle = \frac{1}{\sqrt{2}}\left|\psi\right\rangle\end{gathered} \tag{3.72}$$

Under the concept developed by S. Doronin, subspaces, like 'unmanifested pictures on an undeveloped film', are initially contained in the multidimensional information field of the initial quantum tribosystem. As a result of decoherence because of an action of the quantum tribometer on the tribosystem, these subspaces materialize in the form of a manifested tribosystem.

For more detailed theoretical studies of the decoherence in a quantum tribosystem, consider the commutator of the generalized coordinate q and the Hamiltonian H of the system. The choice of quantum expression is associated with the fact that it establishes a connection between the main numerical characteristics of physical systems – coordinate, energy and momentum:

$$[qH] = qH - Hq = \frac{i\hbar}{m}p \tag{3.73}$$

where q is the generalized coordinate operator; H is the Hamiltonian; p is the momentum operator; $\hbar$ is the Planck's constant; m is the mass; i is the imaginary unit.

$$\dot{p} = -\frac{i}{\hbar}[pH] \tag{3.74}$$

Calculating the commutator in formula (3.74) that is weakly dependent on time, which is quite natural for the two quantities p and H considered the integrals of motion, we find the antiderivative of the function $\dot{p}$, which is linearly dependent on time. If we relate this antiderivative with a unit of time, then, within a certain constant, combining (3.73) and (3.74), we can get a relation between the commutators (3.73) and (3.74) for the time unit:

$$[qH] = m^{-1}[pH] \tag{3.75}$$

The simple form of (3.75) makes it possible to define the physical essence of the parameter m, from which we can make a conclusion about the form of the matter in quantum state. Using the matrix form of the operators H, p, q (3.20)–(3.21), we get

$$\begin{aligned}
Hp &= \left\| \begin{matrix} 0 & -i/2 & 0 & 0 \\ 3i/2 & 0 & -3i/2 & 0 \\ 0 & 5i/3 & 0 & -5i/\sqrt{2} \\ 0 & 0 & 7\sqrt{3}i/2 & 0 \end{matrix} \right\|, \\
pH &= \left\| \begin{matrix} 0 & -3i/2 & 0 & 0 \\ i/2 & 0 & -5i/2 & 0 \\ 0 & 0 & 0 & -7i/2 \\ 0 & 0 & 5\sqrt{3}i/2 & 0 \end{matrix} \right\| \\
Hq &= \left\| \begin{matrix} 0 & 1/2 & 0 & 0 \\ 3/2 & 0 & 3/\sqrt{2} & 0 \\ 0 & 5/\sqrt{2} & 0 & 5\sqrt{3}/2 \\ 0 & 0 & 7\sqrt{3}/2 & 0 \end{matrix} \right\|, \\
qH &= \left\| \begin{matrix} 0 & 3/2 & 0 & 0 \\ 1/2 & 0 & 5/2 & 0 \\ 0 & 3/\sqrt{2} & 0 & 7\sqrt{3}/2 \\ 0 & 0 & 5\sqrt{3}/2 & 0 \end{matrix} \right\|
\end{aligned} \tag{3.76}$$

Under (3.73), the commutators [pH] and $[qH]$ in the matrix form are

$$\begin{aligned}
[\mathrm{pH}] &= \left\| \begin{matrix} 0 & -i & 0 & 0 \\ -i & 0 & -2i/5 & 0 \\ 0 & -5i/3 & 0 & \sqrt{2}i \\ 0 & 0 & -i & 0 \end{matrix} \right\| \\
[qH] &= \left\| \begin{matrix} 0 & -1 & 0 & 0 \\ -1 & 0 & 1/2 & 0 \\ 0 & -\sqrt{2} & 0 & \sqrt{6} \\ 0 & 0 & -\sqrt{6} & 0 \end{matrix} \right\|
\end{aligned} \tag{3.77}$$

Then, by (3.75),

$$\begin{Vmatrix} 0 & -1 & 0 & 0 \\ -1 & 0 & 1/2 & 0 \\ 0 & -\sqrt{2} & 0 & \sqrt{6} \\ 0 & 0 & -\sqrt{6} & 0 \end{Vmatrix} = m^{-1} \begin{Vmatrix} 0 & -i & 0 & 0 \\ -i & 0 & -2i/5 & 0 \\ 0 & -5i/3 & 0 & -\sqrt{2}i \\ 0 & 0 & -i & 0 \end{Vmatrix} \quad (3.78)$$

Equation (3.78) implies that the mass of the 'components' of properly tribosystem, which is in a non-local quantum state, is an imaginary quantity. The latter cannot be explained in terms of classical physics. However, after the publications of Dirac, Feynman and many other theorists who developed the quantum field theory, this behavior can be explained by virtual form of the particles that make the non-local state of a tribosystem. We realize the conflict of concepts 'particle' and 'non-locality', but a virtual particle is an indivisible part of the tribosystem information field, a little similar to a classical corpuscle.

The imaginary nature of such a value as mass emphasizes the difference between virtual and real particles, which is the location of those material (the 'objective reality' in the sense of philosophy) substances in the phase (Hilbert or configuration) space. If we use momentum, energy or just a metric coordinate as generalized coordinates, which make up the so-called Hamiltonian formalism, the area of existence of ordinary particles will be located on the mass surface defined by a relativistic equation:

$$W^2 - c^2p^2 = m^2c^4 \quad (3.79)$$

where W is the particle energy; c is the speed of light; p is the relativistic momentum of the particle; m is the mass of the particle.

Accordingly, the virtual particles with imaginary mass will be beyond this plane. If we transfer our consideration into four-dimensional Minkowski continuum, the virtual world is in a field of negative time (Fig. 3.5) [45].

Considering the matrix relation (3.75) and ignoring certain 'arithmetic roughness', assume an equality $m^{-1} = -i$. In this case, our system is described by the wave function in the form of a plane wave [37]. The free particle described by the 'plane-wave' ψ-function has the energy $W = \hbar^2\kappa^2/2m$. Accordingly,

$$m^{-1} = \frac{2W}{\hbar^2\kappa^2} \quad (3.80)$$

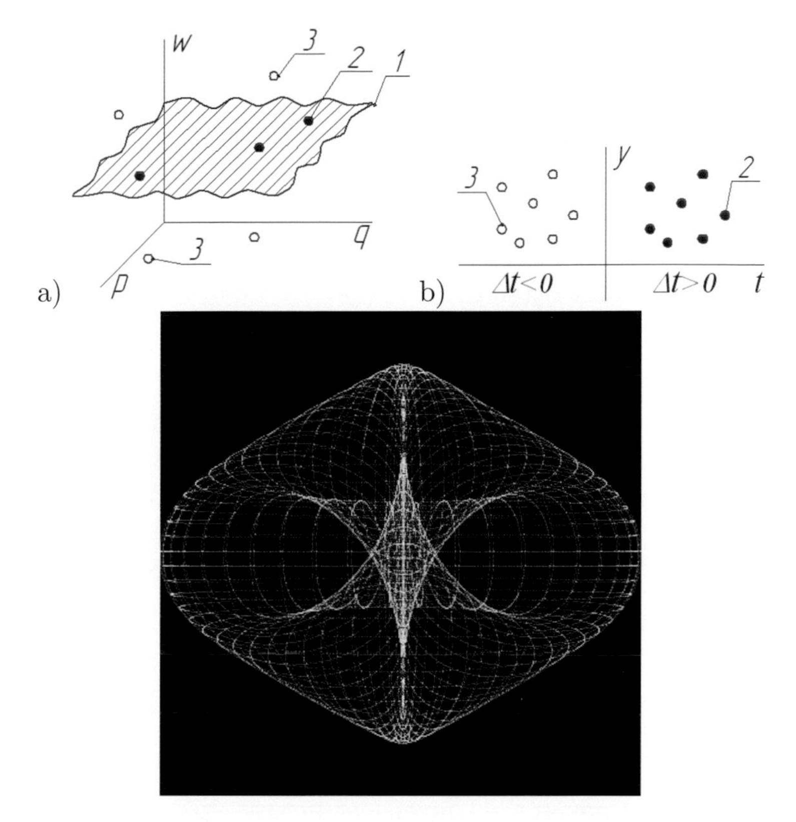

Figure 3.5: Schematic layout of real 2 and virtual 3 particles in the Hamiltonian space (a) with respect to the mass surface 1 and in Minkowski continuum (b); *bottom* – quantum picture

where W is the particle energy; κ is the wave number.

Under the Planck-Einstein formula, $W = \hbar\omega = \hbar c/\lambda$ (within 2π), (3.80) is transformed as

$$m^{-1} = \frac{2c}{\hbar\kappa} \tag{3.81}$$

The wave function ψ that describes the behavior of a free particle as a plane wave has the form $\psi_0 e^{-i\kappa x}$, then

$$\begin{gathered} \frac{\psi_0}{\psi} = e^{i\kappa x} \\ \kappa = \frac{1}{ix} \ln \frac{\psi_0}{\psi} \end{gathered} \tag{3.82}$$

Taking into account the obtained Eq. (3.82), the formula (3.81) is rewritten as

$$m^{-1} = i\frac{2c}{\hbar}\left(\ln\frac{\psi_0}{\psi}\right)^{-1} x \tag{3.83}$$

Taking into consideration the equality $m^{-1} = -i$, (3.83) for the wave function will be written as

$$\psi = \psi_0 e^{-\frac{2c}{\hbar}x} \tag{3.84}$$

In (3.84) the 'damping coefficient' of the wave function is constructed by fundamental physical constants and reaches incredibly big values, approximately equal to 10^{42}. This suggests a virtually immediate wave damping at distances that are multiples of the Planck length [53]. So, the particle in virtual state virtually cannot exist beyond its localization area.

Digression 3.1. Such a big value of the exponent in (3.84) is close to the familiar number 10^{40} obtained by Eddington and Dirac in revealing the numerical relations between the physical parameters of macro- and microworlds. The physicists discovered an amazing regularity that reflects all the relations in the Universe structure, still hidden from us. The number 10^{40} or close to it is the number of atomic nucleus density and the average density of the matter in the universe, Coulomb and gravitational forces, the constants of strong and gravitational interactions, etc. [53].

Chapter 4

Quantum Dynamics of Tribosystems

As suggested in Chapter 3, quantum mechanics as a fundamental concept of the universe is a completely self-sufficient theory, the physical and mathematical formalism of which is sufficient for a science-based description of the birth of material bodies 'out of nothing' and for an explanation of the extraordinary diversity of the controversial picture of frictional interaction. A concrete realization of a tribosystem, the 'manifested tribosystem', is observed in terms of quantum theory, in full accordance with the meaning given to this concept by P. Dirac.

In fact, a manifested tribosystem is a result of the information field perturbation, which is properly tribosystem being in the initial non-local quantum state. A question may arise: where did this state in the friction unit come from? In fact, such a question is incorrect when analyzed more deeply. Under quantum field theory (according to R. Feynman; also quantum matter theory), the non-local quantum state is primary with respect to the real-valued form of the matter, which, as said the American physicist Davy, 'just a trick of information'.

Under the quantum approach, a tribosystem is born by perturbations of the primary information field at the measurement, like splashes from a liquid surface when various items fall there and like those splashes break up into separate drops, individual visible tribosystems arise (Fig. 4.1).

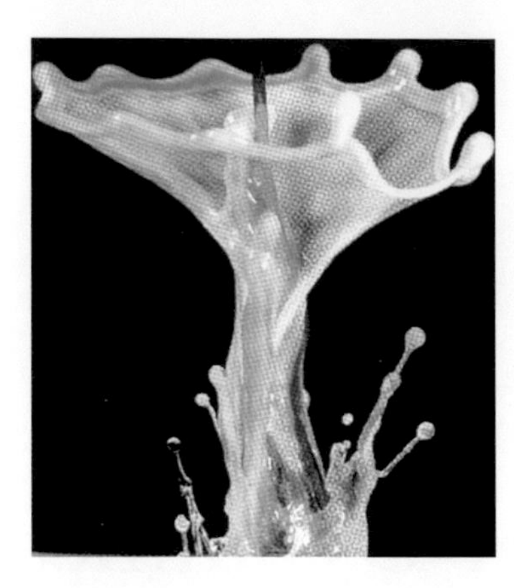

Figure 4.1: 'Dripping' analogy of the quantum birth of a physical object out of the primary information field

Undoubtedly, these analogies are very relative, since quantum objects cannot be described using any classic images. This was stated by the great Soviet physicist and Nobel Prize-winner, L.D. Landau who argued that the biggest victory of the scientific thinking is the ability to describe the processes with it, the visual representation of which is simply impossible. Actually, such items include our quantum tribosystem model, which can be judged only by its manifestations in the four-dimensional Minkowski world.

4.1 Minkowski Space as the Observable Tribosystem World

Under Def. 3.28, what appears to tribology is the result of the decoherence processes that cause the transition of matter from the non-local state into its real classical forms. Outstanding advocates of the 'non-local' may and have the right to criticize this thesis. In some ways, the emergence of real forms of tribosystems may be regarded as a transition of the system from the virtual state (the imaginary mass in (3.78)) into the real state at energy excitation of the physical vacuum. This process is illustrated using the perturbation theory [31], when the probability of the transition is always proportional to the square of the absolute value of the matrix element of perturbation energy relevant to the initial and final states. From the latter, we need to learn to estivate the probability of the transition of a quantum system into a particular state.

Definition 4.1. The probability to find some accurate set of values of a physical quantity ξ observed in its measurement performed on a system in an arbitrary state $\left|x\right\rangle$ is

$$V^{R}_{\xi/x} = \left\langle x \middle| P_{\xi} \middle| x \right\rangle \tag{4.1}$$

where P_{ξ} is the project of this physical quantity (3.57).

The tribosystem materialization that occurs due to the transition from the configuration space where properly tribosystem is in the non-local quantum state into the four-dimensional space-time continuum requires a change in 'quantum rhetoric'. For example, if earlier, in writing the parameters characterizing the quantum system, we used the term 'state vector' in describing the properties of a 'materialized' quantum system, 'state vector' will be replaced by the concept of wave function [28]. In most courses of quantum mechanics, the concepts of wave vector, state vector, wave function and probability amplitude are synonyms. For our case, a fundamental change in the nature of a physical system due to interspatial transitions, we should agree with S.I. Doronin who stated that some difference between the physical concepts of 'state vector' and 'wave function' still exists.

Since the wave function depends on coordinates and time (the existence of the space-time continuum), this already makes the concept of state vector a more general physical quantity. S. Doronin wrote that the wave function is a quasi-classical quantity because in axioms of quantum mechanics there is no such concept as space or time and different space-time continua arise only as a consequence of the decoherence of the 'non-local source of reality'. That is why the properties of the 'observable', the manifested tribosystem, are determined by the peculiarities of the space metric where a particular tribosystem 'incarnation' arises.

The space in which the manifested tribosystem is localized refers to the four-dimensional Minkowski world. Therefore, it has a number of common characteristics and properties inherent in this type of spaces.

The main structural unit of the Minkowski space is an 'event', similar to a point in Euclidean space. The event is characterized by a place 'where' and the time 'when' it occurs. The place and time can be specified only with respect to an inertial frame of reference K where events are specified by four numbers (x, y, z), the common coordinates, and t, the time, together called the event coordinates. At another point of the frame of reference K, the same event has coordinates (x', y', z', t'). Then, under Einstein's postulates (Sec. 1.4, postulates **1** and **2**), light propagation occurs in both systems with the same speed c. Accordingly, we may write the following expressions of analytic geometry:

$$c(t_2 - t_1) = \sqrt{(x_2 - x_1)^2 + (y_2 - y_1)^2 + (z_2 - z_1)^2}$$
$$c(t'_2 - t'_1) = \sqrt{(x'_2 - x'_1)^2 + (y'_2 - y'_1)^2 + (z'_2 - z'_1)^2} \tag{4.2}$$

Equation (4.2) may be represented in geometric form if we assign to each pair of coordinates a number I called interval:

$$-I_{21}^2 = (x_2 - x_1)^2 + (y_2 - y_1)^2 + (z_2 - z_1)^2 - c^2(t_2 - t_1)^2 \tag{4.3}$$

The fact that the events in emission and registration of the same light signal may be written for any coordinate system in the form of the condition $I_{21} = 0$ for the system K and $I'_{21} = 0$ for the system K'. Hence, the interval vanishing does not depend on the coordinate system. As Minkowski wrote, '...zero intervals are invariants' [49]. It can be shown that not only zero, but also small, as well and finite intervals are invariants [55], which allow us to write an equality $I_{21}^2 = (I'_{21})^2$. Accordingly, it is advisable to allow the equality of the square of the distance between the closest events in Minkowski space corresponding to the interval squared. That is why the Minkowski world transforms from an ordinary three-dimensional Euclidean space into a unified (that is why it is so called) four-dimensional space-time continuum.

The metric of the Minkowski world contains only the squares of the coordinate differences (4.2), but one of them is included in an interval with a plus sign and the others with a minus sign (4.3). This space is called pseudo-Euclidean with the signature plus, minus, minus, minus. Due to indefiniteness of the metric of pseudo-Euclidean space, its geometry is more 'diverse' than Euclidean geometry. While a pair of points in Euclidean space (provided they do not match) can be in only one relation to each other, the square of pseudo-Euclidean space can be both positive and negative or zero, which leaves plenty of room for various physical interpretations of such variants, including diverse course of time and change in space curvature [23].

Interesting is the motion representation itself and perhaps its 'non-quantum' characteristic, the path (trajectory). In Minkowski space, to emphasize the temporal basis of all the processes in it, such terms as 'history' and 'world line' are introduced [23].

Definition 4.2. *World line* is a curve in space-time that shows the motion of a classical point particle, as well as the propagation of light beams; it is a continuous sequence of events corresponding to the particle position in space at each moment of time [49].

The term 'history' as a synonym of world line is designed to describe the motion in Minkowski space as some evolutionary process [49]. The light pulse propagating from its source in all directions is a sphere of an increasing radius. In flat projection, this will remind circles on water that radiate if a stone is thrown in it. If we imagine the three-dimensional model of Minkowski continuum, of which two dimensions are the surface of water and the third one is the time axis, in this model a circle radiating in water appears as a cone with the vertex at the point of the stone dropped into water (Fig. 4.2). Similarly a light cone is formed that divides four-dimensional Minkowski space into the past and the future (Fig. 4.3) [49].

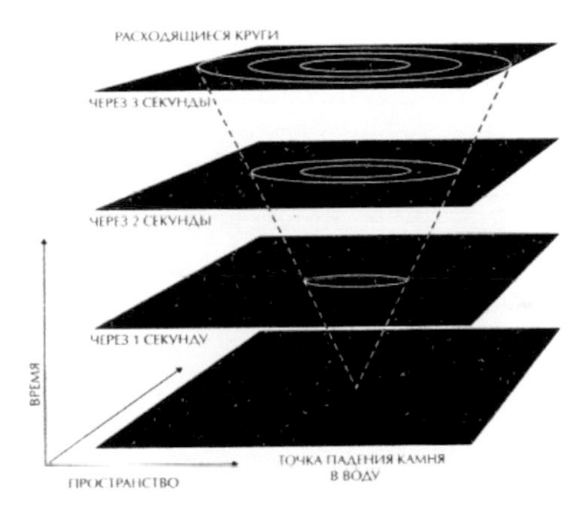

Figure 4.2: The Minkowski space-time in the model of 'a stone dropped into water'

The world line of free particles is a straight line, which, to some extent, reflects the execution of the law of inertia. Meanwhile the impact of the forces contorts its form [23], thus indicating the impact of the force fields on the metric of the Minkowski continuum. It is worth remembering that, under the basic regulations of relativistic physics, space-time does not exist apart from the material world, it 'appears' and 'disappears' together with it. The peculiarities of space-time do not depend on the parameters of the material bodies that compose it.

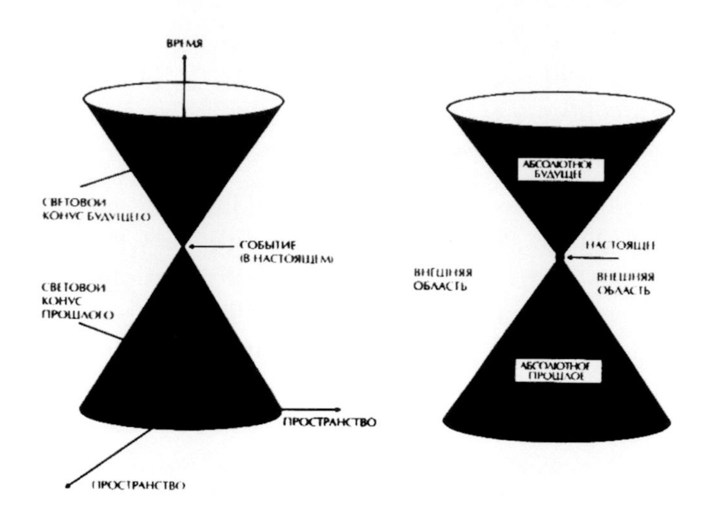

Figure 4.3: The light cone formed in Minkowski space-time

For example, in general theory of relativity, Albert Einstein removed gravitation as a physical force and replaced this physical concept by the geometric properties of space-time continuum. These properties change (a curvature of world lines occurs) under the influence of various forms of energy, including the direct impact of massive objects on space. In this sense, the world line of the universe may be regarded in terms of the famous American mathematician M. Gardner, "...like a monstrous ball of tangled twine with billions on billions of crossings, the 'string' filling the entire cosmos in one blinding, timeless instant. If we take a cross section through cosmic space-time, cutting at right angles to the time axis, we get a picture of 3-space at one instant of time. This three-dimensional cross section moves forward along the time axis, and it is on this moving section of 'now' that the events of the world execute their dance" [56]. The motion ('history') looks very interesting in the Minkowski world. His great disciple, Albert Einstein argued that the past, the present and the future are nothing but a human illusion devoid of physical meaning. Therefore, the past has not disappeared, the future is as real as the present. All things freeze forever on the Minkowski-Einstein world events line. The world line becomes a 'line of fate of the world' [23].

The manifested tribosystem that appeared because of the decoherence creates an event space for itself, four-dimensional Minkowski continuum, which before the act of measurement followed by the decoherence and the emergence of local states simply did not exist. An event of 'Everett kind' occurred: new universe is born. Owing to quantum, as well as classical laws of relativistic physics, the space-time metric is determined by the state of the manifested tribosystem as a whole. At the same time, wide enough opportunities of the pseudo-Euclidean Minkowski world allow various impacts of 'relativistic kind' on the internal processes of the manifested tribosystem [57].

Now imagine the motion of a classical 'manifested' physical system in the Minkowski continuum, which may be described by the Lagrange equation [58]:

$$\frac{d}{dt}\left(\frac{\partial L}{\partial \dot{q}}\right) - \frac{\partial L}{\partial q} = Q \tag{4.4}$$

where $L = L(q\dot{q}t)$ is the Lagrange function (the Lagrangian); q are generalized coordinates; $\dot{q}$ are generalized vectors; t is time; Q are generalized forces.

The force has field nature defined by potential energy U:

$$Q = -\frac{\partial U}{\partial q} \tag{4.5}$$

With the use of (4.5), the Lagrange equation looks like

$$\frac{d}{dt}\left(\frac{\partial L}{\partial \dot{q}}\right) - \frac{\partial L}{\partial q} = -\frac{\partial U}{\partial q}$$

(4.5) has a significant restriction: it is valid only for conservative forces. Therefore, in general case, the generalized force parameter is reasonable to be defined with the generalized potential Φ [52]:

$$Q = \frac{\partial \Phi}{\partial q} - \frac{d}{dt}\left(\frac{\partial \Phi}{\partial \dot{q}}\right) \tag{4.6}$$

For this case, the Lagrange equation (4.4), if the Lagrangian equality $L = W_\kappa - U$ (where W_κ is kinetic energy) is met, is rewritten as

$$\frac{d}{dt}\frac{\partial}{\partial \dot{q}}[W_\kappa - (U - \Phi)] = \frac{\partial}{\partial q}[W_\kappa - (U - \Phi)]$$

or (4.7)

$$\frac{d}{dt}\frac{\partial W_\kappa}{\partial \dot{q}} = \frac{\partial}{\partial q}[(\Phi - U)]$$

Knowing that $W_\kappa = a\dot{q}^2$ in generalized coordinates, we may rewrite:

$$\ddot{q} = \frac{0.5}{a}\left[\frac{\partial}{\partial q}(\Phi - U)\right] \tag{4.8}$$

The quantity $\frac{\partial}{\partial q}(\Phi - U)$ is a 'generalized' force created by a vector force field and its relation to the parameter a in our system may be interpreted as a source (or one of the sources) that creates this force field. Then $a^{-1}\left[\frac{\partial}{\partial q}(\Phi - U)\right]$ is the intensity of the field G. So (4.8) may be rewritten as

$$\ddot{q} = 0.5G \tag{4.9}$$

It turns out that, up to a constant, the acceleration and the field are the same quantity, which matches the geometric interpretation of gravitation that relates the latter with the space curvature in Einstein's general theory of relativity. It clarifies the existing impact on the metric of the continuum where the tribosystem is localized and the source of which are dissipative energy flows generated by friction forces. The energy density of these streams determines the 'depth' of this impact: the Euclidean nature of the properties of the tribosystem continuum retains a negligible effect. When the energy density values are large, the space curves and becomes non-Euclidean. With more impact on the internal processes, such effects begin to have a change in flow of time, an increase in mass, called triborelativistic [57, 59]. In this case, it is important that experimental data are currently available affirming the presence of 'relativistic forces' in frictional contact area [60, 61].

Digression 4.1. Since ancient natural metaphysical philosophers, space and time were considered almost 'mathematical abstractions', identical always and everywhere, not depending on the bodies in it. The German mathematician, B. Riemann developed a mathematical formalism for the analysis of spaces that have diverse metrics. In his mathematical models, the space could be bent and twisted, might have gaps and holes and be multidimensional. The fact that it is possible in the real world was proven by numerous works in general theory of relativity. In particular, the famous German physicist and mathematician, K. Gödel wrote the solutions for the basic general relativity equations, in which the world lines of the Minkowski continuum can be closed in themselves, forming loops. Those closed world lines are called Gödel time loops. An observer, being in such space-time 'construction', may well begin a journey through time, falling into the past or the future. In the universe, there are indeed objects, in the vicinity of which the space-time closes in itself forming Gödel loops. These objects are black holes [53]. The famous modern physicist Stephen Hawking wrote: "Even without using the Einstein equations, I was able to show that, in general, a finitely generated Cauchy horizon will contain a closed light ray, or a light ray that keeps coming back to the same point over and over again." It looks like a *déjà vu*. "Moreover, each time the light comes around, it will be more and more blue-shifted, so the images will get bluer and bluer. The light rays may get defocused sufficiently each time round so that the energy of light doesn't build up and become infinite. However, the blue shift will mean that a particle of light will have only a finite history, as defined by its own measure of time, even though it goes round and round in a finite region and does not hit a curvature singularity" (Fig. 4.4) [62].

Figure 4.4: Curvature of light rays by Sun gravitation

Digression 4.2. It is worth clarifying that the words of S. Hawking refer to a black hole whose geometric dimensions are given by the so-called 'Cauchy horizon', a term introduced into modern physics by Hawking and Penrose. It means a boundary of physical phenomena causal predictability in the future from their initial conditions; a change in the color (spectral composition) of light radiation is a manifestation of the Doppler effect caused by the change in the flow of time.

The relationship of distortion of space-time continuum and energy flows inevitably leads to the fact that under heterogeneous energy distribution, the spatial 'heterogeneity' arises. It can probably be represented as separate subspaces of four-dimensional Minkowski continuum. This variant corresponds to the manifested tribosystem, which is an actively evolving physical object. This tribosystem evolution is treated as a sequence of events of birth, change and death of its event space.

The distortions of space-time continuum of the manifested tribosystem can reach such high values that not only the 'red', but also the 'blue' Doppler shift of the laser beam passing through the friction area occurs. Detailed descriptions of the experiments on registration of the Doppler shift by friction can be found in (Fig. 4.5) [60, 61]. The shift of laser radiation to the short-wave spectral range means the appearance of closed world lines and the related histories in the tribosystem spatial metric, and consequently, the special 'super-dense' energy areas causing such distortions.

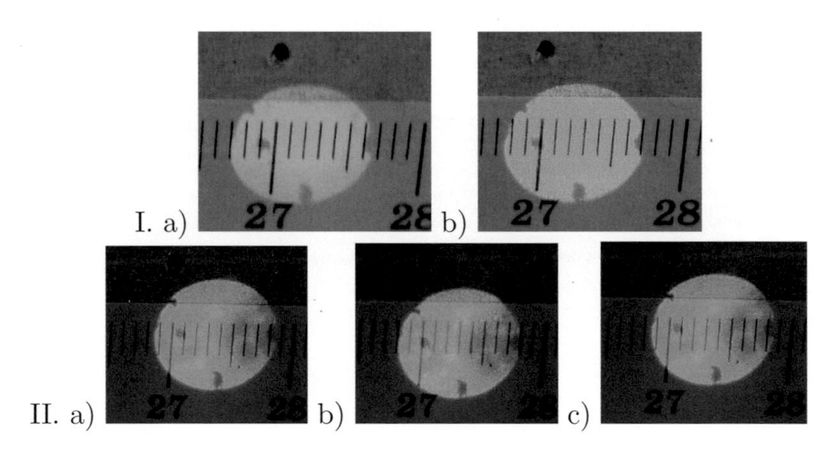

Figure 4.5: Location of light spots on the measuring scale: I – 'red' shift registration by friction: (a) at the beginning of the testing, (b) in 4.5 hours of the tribometer work (laser spot right shift on the measuring scale). II – 'blue' shift registration by magnetic field exposure on the friction unit: (a) before friction; (b) 'blue shift' of refracted laser beam under the influence of magnetic field in friction; (c) disappearance of the 'blue' shift effect in the reverse direction of the magnetic field.

The decoherence of 'properly tribosystem' and its transformation into a 'manifested tribosystem' is a transition from the non-local quantum state of the multidimensional configuration space into a 'material form' located in four-dimensional Minkowski world. In terms of formal mathematical formalism, with this transformation, the state vector initially describing the quantum ensemble non-local state transforms into its partial form, the wave function, which corresponds to only one of the possible representations of the state vector. The wave function describes physical processes with the maximum degree of approximation to the methods of classical physics. Naturally, a part of the information field will be lost. This process of 'information loss', which is probably due to decoherence, has quite a visible counterpart. This counterpart is a geometric operation of volume body projection on a plane.

From descriptive geometry, it is known that projections of objects often distort their original appearance [63]. To confirm this fact, it is enough to consider one's own shadow and see how one's silhouette looks changed. Consider a projection of a sphere on a plane (Fig. 4.6). The circle that is a most adequate representation of the sphere manifests itself only in the case of orthogonal projection. In all other cases, the projection of the sphere has an elliptical shape. And in all projection cases, the information about the original body volume is completely lost. This is identical to the loss of a number of dimensions and the information related. Therefore, it should be expected that on transition from the non-local quantum multidimensional form of original tribosystem being into its observable state, a significant part of the information would be lost. So it is not possible to get detailed information about properly tribosystem by separate, observable manifested tribosystems.

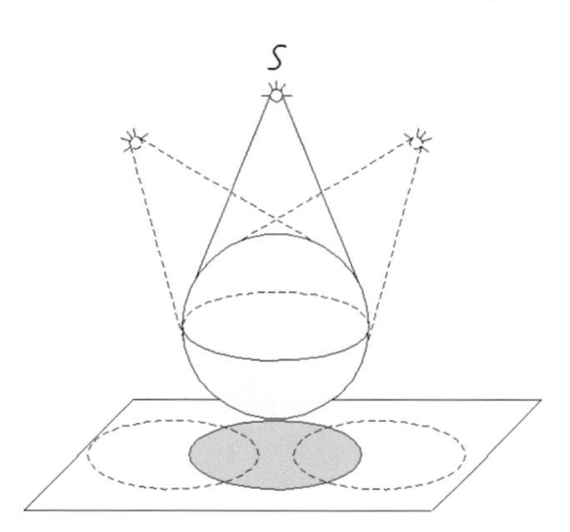

Figure 4.6: Image distortion associated with projection of a three-dimensional body on a plane

In geometry there is a quantitative characteristic of the information loss associated with spatial transitions, which undoubtedly refers to the projection operation. This characteristic is called the coefficient of image incompleteness K_f. It consists of the coefficient of image incompleteness of one parameter of the projecting lines and the three parameters of the plane projection [63]. Considering physical processes and their relationship with the information loss that occurs as a result of spatial transitions requires considering the inevitable symmetry breakdowns and the changes in physical processes caused by them. Actually the act of measurement that causes the reduction of wave function got quite a clear geometric interpretation as projections of a multidimensional configuration object (realizing all the conventionality of the term 'object' in relation to the information field) onto four-dimensional Minkowski continuum. This corresponds to the transformation of the complex parameters that describe the state of the initial object in their real average values. The presence of statistical weight in (2.16) to calculate the average value of a physical quantity leads to the act of measuring a physical quantity to the only possible reality, that is, the one projection of initial non-local physical system 'generated' by the measurement. The frictional parameters, which are the quantitative characteristics of the manifested tribosystem, are the functions of 'histories'. Depending on the specific type of measurement, these are either mechanical or physical and chemical value.

The manifestation of the tribosystem that has its own space-time continuum, resulting from the decoherence, suggests that this space has not only its own metric, but also its own local time, which determines the zero point of the coordinate system and the moment called 'the present'. The category of time, as one of the equal continuum parameters of the manifested tribosystem existence, is important in connection with the transition to quasi-classical methods of describing the local quantum states. The central concept in this description is the concept of motion whose physical meaning is much wider than that declared in the course of general physics on Galileo-Newton mechanics.

The concept of motion characterizes any change that occurs in all systems of the Universe as a whole. In this sense, the motion is rather a philosophical category that is a mode of existence of the entire material world [25], and, in this case, the motion is manifested as a characteristic of any development path, including both the progress and the regress. Here, time is the only parameter that allows to identify the development direction of the objective world. In most theories, time appears as a unity, which is a kind of 'cross-country on which both the past and the future are located' [29]. This view of the nature of time completely excludes the existence of a special moment called the present (unless, of course, it coincides with the origin), and therefore, the allocation into something special of the sequence of a number of transformations of future events into the present and then into the past. This is reflected in the time reversibility of the basic laws of physics. In its terms, it does not matter for Nature whether a time parameter with a 'plus' sign is included in physics equations, which tells about moving from the past to the future, or with a 'minus' sign, which indicates the time reverse. The many-world (Everett, in fact) representation of tribosystem evolution developed in this book allows the existence of various space 'structures' with a lot of generalized coordinates and, accordingly, with a variety of time directions. Between the events at different space-time points a quantum connection should exist based on the non-local nature of quantum laws (the entanglement of quantum systems), which ultimately limits thc number and kinds of variants of the observable reality manifesting itself in the form of manifested tribosystems.

4.2 Quantum Principle of Least Action (Feynman Diagrams and Tribosystem Wave Function)

In accordance with the above, each manifested tribosystem that arises from the decoherence of space-time continuum can have its own, distinct from other, metric defined by energy dense in this space. The inevitable changes that take place in this space may be considered from the standpoint of energy evolution in which a change in the total amount of energy occurs, e.g. due to the transition and the energy quality in dissipative processes [52].

Previously, we defined any quantitative and qualitative change as motion. We supplement this phenomenon by a more physical definition.

Definition 4.3. Any energy evolution of a physical system, i.e. any process of radiation or absorption of energy by a physical body or a group of bodies integrated in a system, may be considered as motion [25].

For the numerical description of 'energy' interpretation of a motion process as the main characteristic, the concept of 'action' is used.

Definition 4.4. Action S is a fundamental physical quantity whose specification as a function of the variables describing the system state completely defines the system dynamics; action for a time interval (t_1, t_2) may be expressed by the Lagrange function $L(q, \dot{q}, t)$ (4.4):

$$S = \int_{t_1}^{t_2} L(q, \dot{q}, t)dt$$

or by Hamiltonian (4.10)

$$S = \int_{t_1}^{t_2} \left[\sum_i p_i q_i - H(p, q, t)dt \right]$$

Under the physical meaning of the Lagrangian as the difference of kinetic and potential energy, action may be written as

$$S = \int_{t_1}^{t_2} (W_\kappa - U)dt \tag{4.11}$$

Motion is subject to the principle of least action.

Definition 4.5. The *principle of least action* is formulated as follows: for each physical system, there exists a quantity called action (4.10, 4.11) that takes the least value when the motion is really happening [58].

It follows from the principle of least action:

Corollary 4.1. Minimizing the action provides real motion only on those paths for which at a given time the difference of kinetic and potential energy is minimal, i.e. $(W_\kappa - U)\Delta t \to \min$.

In modern theoretical physics, action is considered a fundamental quantity when any theory is formulated whose equations are derived from variational principles of mechanics. With this in mind, the theory is formulated as a problem of choosing the metric of the phase space [52] where the state of physical system is defined. The action rate of change is defined by an operator of the Hamiltonian of the physical system. Extremely important is the role of action in systems described by methods of quantum mechanics.

If the basis of a quantum system space is formed by proper operators of state $\left|q\right\rangle$, the transition from space $\left|q_1\right\rangle$ to space $\left|q_2\right\rangle$, under 3.4 is defined by amplitude $\left\langle q_2(t_2)\middle|q_1(t_1)\right\rangle$. Thus, the transition is complete in time interval (t_1, t_2):

$$\left\langle q_2\middle|q_1\right\rangle = \int\limits_q \prod dq(t) \exp\left[-\frac{i}{\hbar}\int\limits_{t_1}^{t_2} L(q,\dot{q},t)dt\right] \tag{4.12}$$

where $\prod$ is a product showing that the exponent integration of classical action is taken over all possible paths beginning at the moment (q_1, t_1) and ending at (q_2, t_2).

The formulation of the principle of least action in the form (4.12) additionally clarifies the nuances of transition from quantum to classical description of physical processes. In quasi-classical limit, for which it can be assumed that a quantum of action $\hbar$ tends to zero and the values of the phase $S/\hbar$ are high, the stationary domain S (where $S = \text{const}$) will make the main contribution to the integral (4.12). Therefore, the action variation δS will be equal to zero. Then the principle of least action itself is a consequence of quantum dynamic laws in a quasi-classical limit. In a certain meaning, the concept of action is more important for quantum theory than for classical physics. For quantum dynamics, under (4.12), all the possible 'paths' are important, with only the extreme ones for the classical principle of least action.

Note 4.1. The concept of path (trajectory) is purely classical and in connection with the uncertainty principle makes no sense for quantum-mechanical objects. In this case, path is used as a characteristic of the system history in a certain phase space, in this sense the term 'path (trajectory)' contains a deeper meaning compared to 'line that draws a body in space over time'. In this meaning, that reflects the history of the evolutionary process of a physical object, the term 'path (trajectory)' is quite widely used in quantum mechanics as well; in fact, there is a rational justification of Feynman's sum-over-histories principle [54].

The 'sum-over-histories' approach has become one of the main tools of quantum field theory. Stephen Hawking estimated the value of Feynman's work: "Feynman's theory gives an especially clear picture of how a Newtonian world picture can arise from quantum physics, which seems very different. According to Feynman's theory, the phases associated with each path depend upon Planck's constant. The theory dictates that because Planck's constant is so small, when you add the contribution from paths that are close to each other the phases normally vary wildly, and so... they tend to add to zero (Fig. 4.7a). But the theory also shows that there are certain paths for which the phases have a tendency to line up, and so those paths are favored; that is, they make a larger contribution to the observed behavior of the particle (Fig. 4.7b). It turns out that for large objects, paths very similar to the path predicted by Newton's will have similar phases and add up to give by far the largest contribution to the sum, and so the only destination that has a probability effectively greater than zero is the destination predicted by Newtonian theory, and that destination has a probability that is very nearly one... Feynman's theory allows us to predict the probable outcomes of a 'system', which could be a particle, a set of particles, or even the entire universe. Between the initial state of the system and our later measurement of its properties, those properties evolve in some way, which physicists call the system's history... Feynman showed that, for a general system, the probability of any observation is constructed from all the possible histories that could have led to that observation. Because of that, his method is called the 'sum over histories' or 'alternative histories' formulation of quantum physics" [27].

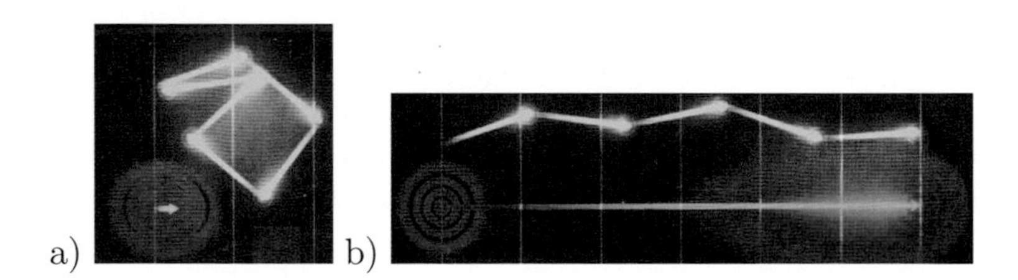

Figure 4.7: Adding up Feynman's paths: (a) sum of the paths is very short; (b) sum of the paths is long (the yellow arrows are the phases summed; the blue arrows are the sum of the paths from the tail of the first arrow to the point of the last one).

And here is how Feynman himself explains his method by an example of light propagation: "For each crooked path, such as path A (Fig. 4.8), there's a nearby path that's a little bit straighter and distinctly shorter – that is, it takes much less time. But where the paths become nearly straight – at C, for example, a nearby, straighter path has nearly the same time. That's where the arrows *(Feynman means state vectors—author's)* add up rather than cancel out; that's where the light goes... the single arrow that represents the straight-line path, through D (Fig. 4.8) is not enough to account for the probability that light gets from the source to the detector. The nearby, nearly straight paths – through C and E, for example – also make important contributions. So light doesn't *really* travel only in a straight line; it 'smells' the neighboring paths around it and uses a small core of nearby space. (In the same way, a mirror has to have enough size to reflect normally: if the mirror is too small for the core of neighboring paths, the light scatters in many directions, no matter where you put the mirror)" [54].

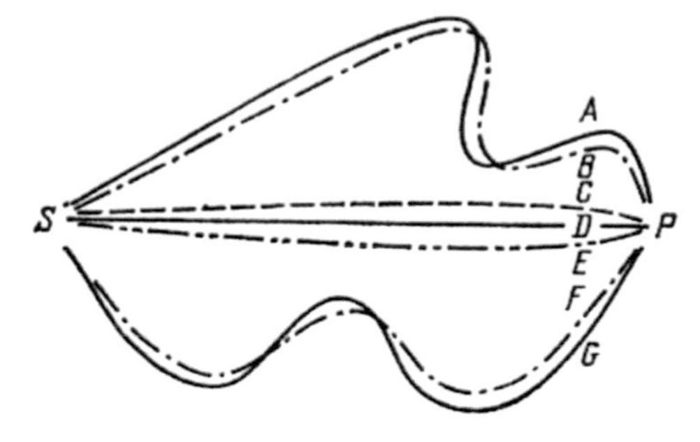

Figure 4.8: The method of adding up quantum 'histories' [54]

Feynman warns of searching for simple 'mechanical' similarities of this quantum systems behavior. He wrote: "There are no 'wheels and gears' beneath this analysis of Nature; if you want to understand Her *(Nature—author's)*, this is what you have to take [54]".

Together with the creation of a quantum model of alternatives histories, Feynman developed a beautiful and concise graphical method of accounting the different stories – Feynman diagrams [27, 54]. The effectiveness of this method was noted by another eminent physicist, Nobel Prize winner, L. Cooper who wrote that Feynman's diagrams "...instead of thousand words made easier (if not possible at all) the life of the modern generation of physicists and made a substantial simplification in the calculations of 'bookkeeping' type" [64]. The sum over all possible histories can be represented as a sum over Feynman diagrams. They represent the process of birth and evolution of a quantum object at a definite point of space-time continuum. Since the three directions in space have the same properties, the three spatial coordinates are replaced by one (e.g. x). The y-axis will be the time. Thus, four-dimensional Minkowski space is converted into a two-dimensional space of quantum particles interaction denoted by a point with coordinates (x, t), called vertex. The motion of real particles on Feynman diagrams (we will again use the term 'graphs') are shown by a solid line, virtual particles – by an intermittent line, and photons – by a wavy line [54].

Feynman proposed to consider antiparticles as the particles that move backwards in time. He wrote: "Even more strange is the possibility that the electron emits a photon, then *travels backwards in time* to absorb a photon, and then proceeds forward in time again. The path of such a 'backward-moving' electron can be so long as to appear real in an actual physical experiment in the laboratory... The backward-moving electron when viewed with time moving forwards appears the same as an ordinary electron, except it's attracted to normal electrons – we say it has a 'positive charge'... For this reason it's called a 'positron'... This phenomenon is general. Every particle in Nature has an amplitude to move backwards in time, and therefore has an antiparticle. When a particle and its antiparticle collide, they annihilate each other and form other particles. (For positrons and electrons annihilating, it is usually a photon or two). And what about photons? Photons look exactly the same in all respects when they travel backwards in time... so they are their own antiparticles" [54]. Therefore, at least at the level of quantum systems, the time travels are quite real. The time itself is transformed from a 'sacred cow' into an operating parameter, the same (or nearly the same) as energy or momentum.

To distinguish a particle and an antiparticle on Feynman diagrams, arrows are put on their paths indicating the direction of motion in time. An example is a diagram illustrating the interaction of electrons, positrons and photons (Fig. 4.9). The graphs show the sequence of annihilation events of a particle (electron) and an antiparticle (positron), on which there is initially an interaction of the electron (1) depicted by a solid line with a right arrow (time going forward) and the positron (2) a solid line with a left arrow (time going backwards) (Fig. 4.9). Their interaction, denoted by a vertex point, indicates the photon (3) is born in the course of their annihilation. The reverse process is also possible, in which at some point in time t_1 the photon 'splits' into two particles: an electron and a positron (Fig. 4.9). The positron, not living long, at the moment t_2 (very close to t_1) collides with the nearest electron and their annihilation forms a new photon. Meanwhile, the electron formed from the primary photon keeps moving in Minkowski space [54].

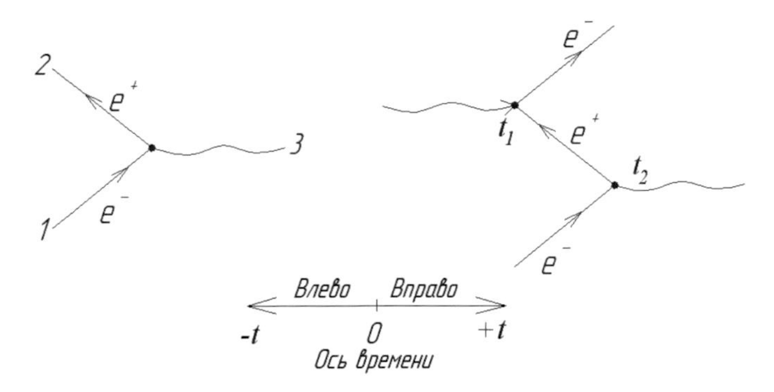

Figure 4.9: Feynman diagrams of electron-positron interaction

Let us consider in more detail the issue of motion 'backwards in time'. The real possibility of such a motion lies in the so-called time reversibility of the basic equations of quantum and classical (including not only the theory of relativity, but also Newtonian physics) mechanics. There replacing the signs at the time parameter t plus by minus does not change the nature of the action of physical laws. In quantum mechanics, this is manifested as the symmetry of the Schrödinger equation solutions with respect to time sign. This means in particular that if a quantum-mechanical motion described by a wave function ψ is possible, then possible is the motion described by a conjugate wave function ψ^*, at which the system passes through the same states but in reverse order. Thus, in quantum mechanics, the symmetry of laws with respect to time direction is observed at transition from wave functions ψ to their conjugates ψ^*. Therefore, the wave function describing the particle motion backwards in time axis coincides with the complex conjugate value of the direct wave function.

In the monograph [37] devoted to quantum mechanisms of frictional interaction, the friction process was considered as an alteration of activation and passivation of tribounits working surfaces. It has been shown that if the activation process is described by a wave function ψT, the passivation, respectively, by $\psi^* T$. Consequently, the surface passivation phenomenon (which undoubtedly matches the actual course of events) may be considered as activation backwards in time axis. Feynman argued that the journey back along the time axis does not have to exactly repeat the original path. The latter can be, e.g. of a zigzag shape (Fig. 4.10), where a return is not to the starting point A but to the point B at a distance of $\Delta r = CD$ in space, which is essentially a chronological twin of the point A [57].

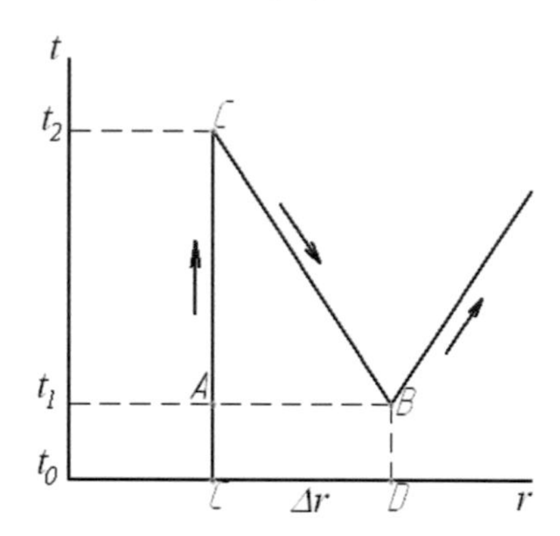

Figure 4.10: Feynman's scheme for time travels

This form of the world line solves the problem of violation of the causality principle formulated, e.g. by analogy with the 'Schrödinger's cat' as follows: "What will happen to the grandson, the time traveler, if he kills his own grandfather before the birth of his father?" If the events are held by the 'scenario' of the path indicated in Fig. 4.10, the violation paradoxes of the causality principle are not relevant. Under the Everett doctrine, grandson, the traveler, as a result of moving back in time, 'perturbs' Minkowski continuum and creates a new parallel reality represented in the diagram at point B. Here he can kill his grandfather and, naturally, the grandson's doppelganger will never appear again in this world, although in the universe where he began his journey, all things remain unchanged. According to M. Gardner, "...time dilation in relativity theory, time travel in Gödel's cosmos, and reversed time in Feynman's way of viewing antiparticles are so carefully hedged by other laws that contradictions cannot arise" [65]. The motion back in time, though sounds fantastic, is not something prohibited in terms of physical laws.

The description of quantum processes using Feynman graphs may be useful with regard to tribosystems, since the history of quantum objects is developing here in Minkowski space, which includes localization continua of post-decoherent states of 'materialized' tribosystems. For this reason, the elements of Feynman's approach were used to build the structure of the quantum model of a tribosystem (Fig. 4.11) [37].

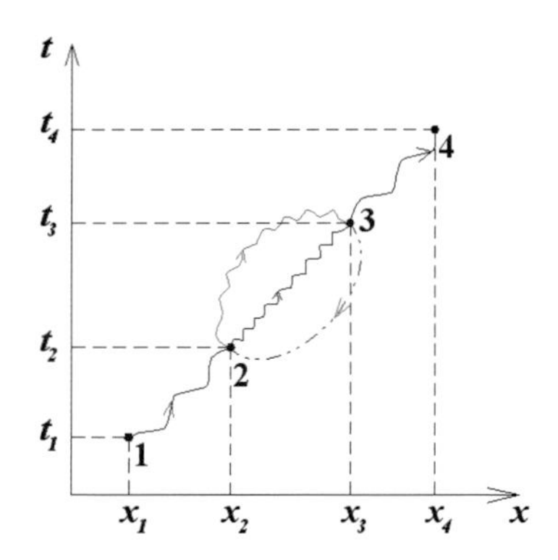

Figure 4.11: Feynman diagram depicting the evolution of a tribosystem in phase space-time $\{x, t\}$, where x is a space coordinate, t is time

In Fig. 4.11, the Feynman diagram of a manifested tribosystem has four clearly marked vertices denoted as $1, 2, 3, 4$. Each of them corresponds to its own development stage of the manifested tribosystem. The tribounit run-in corresponds to vertex 1. At vertices 2 and 3, complex physical and chemical processes are carried out, determining the materials structure adaptability associated with the emergence of new quasi- and virtual quantum particles. The friction established corresponds to vertex 4.

Each vertex in diagram 1 corresponds to the wave function ψ_{ii}, e.g. vertex 1 – ψ_{11}, vertex 2 – ψ_{22}, etc.: ψ_{33}, ψ_{44}.

Under the requirements for this type of quantum processes description, each 'history scenario' of the manifested tribosystem should be characterized by its own wave function: $\psi_1, \psi_2, \psi_3, \psi_4$, the number of which matches both the number of the vertices in the diagram and the number of dimensions in Minkowski continuum. All these history wave functions can be represented in the form of finite functional series, the terms of which will also be the wave functions of the form $\psi_{ii}, \psi_{ij}, \psi_{ji}$, characterizing, respectively, the system states at vertices and the probability amplitudes of the transitions between the vertices:

$$\begin{aligned}
\psi_1 &= \psi_{11} + \psi_{21} + \psi_{31} + \psi_{41} \\
\psi_2 &= \psi_{12} + \psi_{22} + \psi_{32} + \psi_{42} \\
\psi_3 &= \psi_{13} + \psi_{23} + \psi_{33} + \psi_{43} \\
\psi_4 &= \psi_{14} + \psi_{24} + \psi_{34} + \psi_{44}
\end{aligned} \tag{4.13}$$

In (4.13), unlike (2.14), to make the analysis easier, there is no vector expansion coefficient parametrically presented in wave functions ψ_{ij}, ψ_{ji}. Since a single measurement may give only one state and the states are developing 'space-continuously', the probability of crossing the state is negligible, then (4.13) can be simplified to the form (4.14), excluding the wave functions associated with the 'forbidden' transitions:

$$\begin{aligned} \psi_1 &= \psi_{11} + \psi_{21} \\ \psi_2 &= \psi_{12} + \psi_{22} + \psi_{32} \\ \psi_3 &= \psi_{23} + \psi_{33} + \psi_{43} \\ \psi_4 &= \psi_{34} + \psi_{44} \end{aligned} \tag{4.14}$$

According to Feynman's sum-over-histories principle, we introduce a new wave function: $\psi = \sum\limits_{i=1}^{n} \psi_i$

$$\psi = \psi_{11} + \psi_{21} + \psi_{12} + \psi_{22} + \psi_{32} + \psi_{23} + \psi_{33} + \psi_{43} + \psi_{34} + \psi_{44} \tag{4.15}$$

In (4.15) the wave functions ψ_{ii}, ψ_{ij} and ψ_{ji} are presented as terms. It is known that, depending on the laws of quantum statistics (Fermi – Dirac, Bose – Einstein) physical bodies are subject to, their wave function can be asymmetric: $\psi_{ij} = -\psi_{ij}$ or symmetric: $\psi_{ij} = \psi_{ji}$ [24]. Accordingly, the history ψ takes two forms: F for antisymmetric fermions and B for symmetric bosons.

$$\begin{aligned} F &= \psi_{11} + \psi_{22} + \psi_{33} + \psi_{44} \\ B &= F + 2(\psi_{21+\psi_{32}+\psi_{43}}) \end{aligned} \tag{4.16}$$

Note that, along with the transitions described by the wave functions ψ_{21}, ψ_{32}, etc., possible are the states with the wave functions ψ_{12} and ψ_{13}, which characterize (in Feynman's terms) the time reverse transitions, which are defined in the second equation of (4.16) by the coefficient 2. The time reverse transitions in tribosystems are associated only with Bose particles described by the wave function B. The complete history of the tribosystem ψ_R is a sum of B and F:

$$\psi_R = B+F = 2\left[F+\sum_{i=1}^{3}\sum_{j=1}^{4}\psi_{ij}\right] = 2[\psi_{11}+\psi_{22}+\psi_{33}+\psi_{44}+\psi_{21}+\psi_{32}+\psi_{43}] \tag{4.17}$$

Quantum electrodynamics rules [54] allow writing the coupling equations between the wave functions in (4.17):

$$\begin{aligned}
\psi_{21} &= \psi_{11}\Delta(\delta)\\
\psi_{22} &= \alpha\psi_{21} = \alpha\Delta(\delta)\psi_{11}\\
\psi_{32} &= \psi_{22}\Delta(\delta) = \alpha\Delta^2(\delta)\psi_{11}\\
\psi_{33} &= \alpha\psi_{32} = \alpha^2\Delta^2(\delta)\psi_{11}\\
\psi_{43} &= \psi_{33}\Delta(\delta) = \alpha^2\Delta^3(\delta)\psi_{11}\\
\psi_{44} &= \alpha\psi_{43} = \alpha^3\Delta^3(\delta)\psi_{11}
\end{aligned} \tag{4.18}$$

The relations show the changes undergone by the wave function in quantum interactions: when the particles collide, the wave function changes its amplitude by some real coefficient α, and at the transition between the vertices, a phase shift $\Delta(\delta)$ is observed, determined by the final value of the length of time required for this transition [54].

In tribology, the coefficient α is associated with the processes of energy dissipation. Thus it characterizes the decrease in the amplitude of the wave function. Based on the known triboprocesses consequences described in [59, 66], one may neglect the terms that contain higher degrees of the coefficient α. After the appropriate transformations and substitution of Eq. (4.18) into (4.17), the expression for the wave function ψ_R will be written as follows:

$$\begin{aligned}
\psi_R &= 2\psi_{11}\left[1 + \Delta(\delta) + \alpha\Delta(\delta) + \alpha\Delta^2(\delta)\right]\\
&= 2\psi_{11}\left[1 + \Delta(\delta) + \frac{\psi_{22}}{\psi_{11}} + \frac{\psi_{22}}{\psi_{11}}\Delta(\delta)\right]\\
&= 2\left[\psi_{11} + \psi_{22} + \Delta(\delta)(\psi_{11} + \psi_{22})\right]\\
&= 2(\psi_{11} + \psi_{22})[1 + \Delta(\delta)]
\end{aligned} \tag{4.19}$$

If we denote in (4.19) the quantity $2(\psi_{11}+\psi_{22})$ as ψ_0', which is a certain amplitude, and $[1 + \Delta(\delta)]$ is an expansion of the exponent $\exp[\Delta(\delta)]$ into a series, the expression (4.19) for the tribosystem history can be represented as

$$\psi_R = \psi_0' e^{\Delta(\delta)} \tag{4.20}$$

Thus, the wave function that describes the behavior of a tribosystem in Minkowski space is a plane wave. Normally the plane wave is a characteristic of free physical objects on which no external force is applied. Such systems may be subject to unconditional application of all the conservation laws, which, taking into account the dissipative nature of friction units, is quite an unexpected result. The components ψ_{ii} and ψ_{ij} in the expression for the wave function ψ_R, under the laws of wave physics, can also be described by plane sinusoidal waves [67]. If, for example, $\psi_{11} = \psi_0 \cos \omega t$, the action of the phase factor $\Delta(\delta)$ on it will result in a phase shift δ: $\psi_{21} = \psi_{11}\Delta(\delta) = \psi_0 \cos(\omega t + \delta)$. Under the rules of trigonometry, $\psi_{21} = \psi_0 \cos(\omega t + \delta) = \psi_0[\cos \omega t \cdot \cos \delta - \sin \omega t \cdot \sin \delta]$. Considering that the phase shift is small, we get:

$$\psi_{21} = \psi_0 \big[\cos \omega t - \delta \sin \omega t\big] = \psi_0 \cos \omega t - \delta \psi_0 \sin \omega t = \psi_{11} + \frac{\delta}{\omega}\dot{\psi}_{11} \tag{4.21}$$

Also, according to (4.18),

$$\Delta(\delta) = 1 + \frac{\delta}{\omega}\frac{\dot{\psi}_{11}}{\psi_{11}} \tag{4.22}$$

In accordance with (4.22), the exponent ($\exp[\Delta(\delta)]$) is written as $e \cdot \exp\left[\frac{\delta}{\omega}\frac{\dot{\psi}_{11}}{\psi_{11}}\right]$. If we present the plane waves ψ_{11} and $\dot{\psi}_{11}$ as complex, the expression for the exponent is simplified to $\exp(i\delta)$, where i is the imaginary unit. Accordingly, the expression for the wave function ψ_R will look like

$$\psi_R = \psi_0 e^{-i\delta} \tag{4.23}$$

where $\psi_0 = e\psi_0'$ is the new value of the wave function amplitude; e is the base of the natural logarithm.

The representation of the wave function describing the state of the manifested tribosystem as a plane wave (4.23) is of great importance to create a dynamic model of a quantum physical system. First, the form of the function (4.23) ψ_R itself allows to reasonably assume that, despite the nature of friction processes, the forces that determine causal relationships within the tribosystem are conservative (it had been supposed in [37, 45]). Another important factor is close to the value of $5(\psi_{11} + \psi_{22})$ amplitude of the wave function, of which the numerical factor 'five' matches the number of paths that describe the evolutionary processes in the tribosystem. The basic point in (4.23) is its rich informative capacity in relation to dynamic processes, which is due to the reference simplicity of the phase δ with the force characteristics of the physical fields that arise in the friction interaction area of the manifested tribounit. Under the laws of electrodynamics [55],

$$\delta = \frac{q}{\hbar} \int\limits_{L} A dl \tag{4.24}$$

where q is electric charge; $\hbar$ is the Planck's constant; A is the force field vector.

In Minkowski space, A is a four-dimensional, or 4-vector.

Definition 4.6. *4-vector* is a characteristic of Minkowski continuum , of which three coordinates are usual components of vector quantities, e.g. Cartesian coordinates: $\{\vec{i}x, \vec{j}y, \vec{k}z\}$, and the fourth one represents time t. This approach to the description of physical processes in Minkowski space-time is determined by features of its metric, in particular, by (4.3) [55].

4.3 The Action of Force Fields on the Wave Function of a Tribosystem

The importance of external influence on the motion of material bodies was always understood. Under the laws of mechanics, a body gets accelerated, which results in the possibility to change its trajectory, provided there is no coincidence of the force vector with the initial direction of motion of the body. In Newtonian physics, where the space metric, as the space itself, is 'separated' from material bodies [23], the change in trajectory of motion does not lead to 'spatial restructuring'. The later conceptions of the phase space a priori marked the existence of this relationship, which has been proven in relativistic physics. Therefore, identical to Newton's first law, the behavior of the world line from Minkowski world, which curves under external forces, is linked with the metric violation of the continuum studied in detail in general theory of relativity [23].

The physical nature of forces is specifically represented in quantum mechanics. According to the exchange interaction theory, the forces between physical bodies arise in an act of virtual particles exchange. Each type of fundamental forces has its own virtual 'representatives', already open or still remaining hypothetical concepts in science. The 'quantum equivalent' of acceleration, which is a consequence of the action of an external force on a physical body, is the change in the basic state of a quantum system. This concept is close to the interaction between the instrument and the quantum object, this book is based on, with such consequences as the reduction of the wave function and the decoherence of the quantum object, which are properly 'external forces'. In [37], it is shown in a simplified form that an external action is able to change the wave function of a physical system and 'perturb' the energy parameters of a quantum object, the latter can be described using the formalism of perturbation theory [31].

Definition 4.7. If the Schrödinger equation describing the quantum system $H_\psi = W_m \psi_m$ does not have an analytical solution and the Hamiltonian $\hat{H}$ may be written as $H = H_0 + V$ and we know the exact solutions of the Schrödinger equation $H_0 = \bar{\omega}_m \phi_m$, then the solution for the initial equation, but already in the form $(H_0 - V)\psi = W_m \psi_m$, is as follows:

$$\psi_n = \phi_n + \sum_m \frac{\left\langle n \middle| V \middle| m \right\rangle}{\bar{\omega}_n - \bar{\omega}_m} \phi_m + \dots$$
$$W_n = \omega_n + \left\langle n \middle| V \middle| m \right\rangle + \sum_{m \neq n} \frac{\left| \left\langle n \middle| V \middle| m \right\rangle \right|^2}{\bar{\omega}_n - \bar{\omega}_m} + \dots \tag{4.25}$$
$$\left\langle n \middle| V \middle| m \right\rangle = \int_V \phi_n^* \hat{V} \phi_m d\tau$$

where V is the perturbation operator.

Note 4.2. (4.25) illustrates the implementation of the formalism of perturbations theory. Instead of Rayleigh – Schrödinger, another approach is applied in quantum mechanics, called Brillouin – Wigner perturbation theory [68]; in its terms, if the Schrödinger equation has the form $(H_0 + V)\psi_m = W_m \psi_m$ and $\psi_m = \sum_n C_{mn} \phi_n$, then

$$C_{mn}(W_n - W_m^0) = \sum_\kappa C_{\kappa n} V_{m\kappa} \tag{4.26}$$

Assuming $m = n$, we get:

$$W_n - W_m^0 = V_{mm} + C_{mm}^{-1} \sum_\kappa C_{\kappa m} V_{m\kappa} \tag{4.27}$$

Resubstituting (4.26) in the right side of (4.27) and allocating each time the terms $\kappa = m$, we get:

$$W_m^{(n)} = \sum_{s=0}^{n} W_m^{(s)} + \Delta^{(n)}$$

$$W_m^{(s)} = \sum_{\kappa_1} \cdots \sum_{\kappa_s} \frac{V_{m\kappa \dots} V_{m\kappa_s}}{\left(W_m - W_{\kappa_1}^{(0)}\right) \dots \left(W_m - W_{\kappa_s}^{(0)}\right)} \tag{4.28}$$

$$\Delta^{(n)} = C_{mm}^{-1} \sum_{\kappa_1} \cdots \sum_{\kappa_s} \frac{V_{m\kappa_1 \dots} V_{m\kappa_s}}{\left(W_m - W_{\kappa_1}^{(0)}\right) \dots \left(W_m - W_{\kappa_s}^{(0)}\right)} C_{m\kappa_m}$$

Here the matrix notation of operators (3.67) is used.

The formalism of Brillouin – Wigner theory is useful when considering degenerate states in the meaning of Def. 2.25. Then the parameter Δ represents the distance between two nearest levels: $\Delta = W_2 - W_1$, and if we assume that the energy eigenvalues of basic state levels have no perturbation, i.e. $V_{ii} = 0$, the condition [68] is met:

$$W_i = \pm \frac{\left|V_{ij}\right|^2}{\Delta} \tag{4.29}$$

From the linear relationship between the level energy and the perturbation magnitude an important corollary arises.

Corollary 4.2. The position of perturbed levels depends on the perturbation magnitude. Under perturbation, the distance between the nearest levels increases, which is called a repulsion of levels, resulting in removal of the degeneration from the quantum system.

Corollary 4.3. The degenerate quantum level with the main quantum number n, when an external field F is imposed on it, disintegrates by $(2n - 1)$ components; the distance between the extreme sublevels of this disintegration is [68]:

$$\Delta W = 3Fn(n-1) \tag{4.30}$$

The use of formalism and the views of perturbation theory as the basis for quantum system dynamics is based on the definition of motion as energy exchange of a physical system with the environment or neighboring physical systems. The external force may be considered a factor that changes the nature of this exchange. In fact, the force action may be judged by the magnitude of its deviation from the rest (equilibrium) state and possibly by the complete change in the energy and, therefore, spatial structure of the physical system. Let us illustrate this thesis by a tribosystem example.

[37] shows that in the superexcited quantum state, the frictional contact substance has a discrete structure consisting of fifteen shells. Then, under Cor. 4.3, the magnitude of the quantum number n is equal to eight. In accordance with (4.30), we get:

$$\Delta W = 1.68 \cdot 10^2 F = 1.41 \cdot 1.21 \cdot 10^2 F \tag{4.31}$$

Representing the number coefficient as the product $1.41 \cdot 1.21 \cdot 10^2$ allows to represent, in terms of coordinates transformation, the energy impact on a quantum system with 'force' F as a space transition from continuum F to another phase continuum ΔW. As the parameter F, one may use the magnitude of the energy of the force field created by dissipative flows, with the intensity G (4.9). The coefficient 1.41 corresponds to $\cos^{-1}(\phi/2)$ in (3.72), and the number $0.83 \cdot 10^2$ ($0.83 \cdot 10^2$ found as $1/1.21 \cdot 10^2$) determines the portion of dissipative flows energy (and, therefore, the work of friction forces) accumulated by friction surfaces. This accurately matches the results published in [16, 59]. Therefore, within existing assumptions considered in Chapter 3, (4.31) can be defined as a numerical description of the impact of external forces, e.g. friction force, leading to decoherence of the system in Minkowski space, which is orthogonal to the initial Hilbert states space of properly tribosystem. The tribosystem formed is characterized by the quantum number 8 with a number of matrix components (3.68) describing the space of the manifested tribosystem that has 15 relatively isolated shells [37].

Note 4.3. In this system, we mix the concepts that classical physics distinguishes: force, intensity and energy; modern physics does not regard them as something different. These physical quantities are manifestations of the energy structure of matter [28, 29], therefore, such a free use of the terms does not break the physical sense of the arguments and the correctness of the conclusions made on their base.

From a formal point of view, the force impact on a quantum system comes to the changes in the exponent of the wave function (4.23), to which the consideration of this process was properly limited in [37]. The recognition of the impact of dissipative forces, such as friction forces, leads to the necessity to consider the processes of oscillations damping and related changes in their frequencies, under the law $\omega_T = \sqrt{\omega^2 - \beta^2}$ [67], where β is the oscillation amplitude damping index. For those processes, the coefficient α from (4.18) is responsible. In quantum model of friction, the frictional surface interaction is represented as an exchange of 'friction quanta', the tribons [37]. Therefore, dissipative processes may be regarded as tribons dissipation on other groups of particles responsible for various states of the substance of the friction unit, plasmons and phonons [37]. This interaction not only decreases the amplitude of the initial wave function ψ_{11}, but also decreases the magnitude of the energy perturbation transferred by a tribon, which may (in terms of 'ideology' of Note 4.3) be represented as an analog of the mean free path $\bar{x}$ in Minkowski continuum.

The magnitude of particles transfer in a flow x is related with the mean free path $\bar{x}$ and the decrease in particles number $\Delta N = N_1 - N_2$ as [69]:

$$x = \bar{x} \ln \frac{N_1}{N_2} \approx \bar{x} \left[\frac{N_1}{N_2} - 1 - 0.5\left(\frac{N_1}{N_2} - 1\right)^2 \right] \tag{4.32}$$

To write (4.32) in a quasi-quantum form, it is necessary to keep in mind that quantum mechanics studies the behavior of not only individual, but also sets of a big number of particles. Therefore, the wave function is, to some extent, a statistical characteristic of a micro-objects ensemble. The famous Soviet theoretical physicist, V.A. Fock argued that, although the wave function characterizes the state of an individual quantum particle, this is done under predefined conditions, taking into account the environment of the particle considered. Therefore, the wave function should also contain information about the structure of the quantum ensemble, which finally "...reflects the objective possibilities inherent to microparticles to discover definite values of physical quantities" [30].

Let us assign to a flow N a certain quantum ensemble described by a wave function ψ, which is proportional to the square root of N [48]. Then in (4.32) we replace N_1 by ψ_{21}^2, N_2 by ψ_{22}^2. The notation of the wave functions matches the equation system (4.18). From that, (4.32) will be rewritten in the 'quasi-quantum form':

$$x = 0.5\bar{x}\left[\left(\frac{\psi_{21}^2}{\psi_{22}^2}\right)^2 - 1\right] \tag{4.33}$$

or, taking into account $\psi_{22} = \alpha\psi_{21}$:

$$x = 0.5\bar{x}(\alpha^{-4} - 1)$$

In extreme cases, we get $x = \bar{x}$. Accordingly, the 'damping' coefficient α of the wave function is equal to 0.8. Since a manifested tribosystem is considered, it seems logical to correlate the coefficient α with some friction characteristic responsible for dissipative processes, e.g. friction coefficient. Define this relation as linear:

$$\begin{aligned} \alpha &= af \\ \alpha &= b\eta \end{aligned} \tag{4.34}$$

where f, η are the coefficients of external and internal friction; a and b are 'fitting' coefficients.

The oscillations damping determined by dissipative processes under the laws of wave physics [67] results in a change of frequency characteristics of the ψ-function, in terms of the law $\sqrt{\omega^2 - \lambda^2}$, where λ is the damping index. In accordance with the damping oscillation equations, one may assume that $\alpha \sim \exp(-\lambda t) = \exp(-t/\tau) = (1 - t)/\tau$, where τ is the relaxation time [67]. [59] shows the numerical data allowing to analyze quantitatively the process described: $t \sim 10^{-6}$ s, $\tau = 4.3 \cdot 10^{-6}$ s, $\lambda^2 = 0.2 \cdot 10^{14}$ s, $\omega \sim 10^{19}$ s^{-1}. Then the frequency change determined the dissipative triboprocesses $\sqrt{10^{38} - 10^{14}}$ s^{-1} is decreasingly small. Formally, the force impact of the dissipative processes generated by friction forces does not lead to a change in the tribosystem wave function in its exponential part. The decrease in the amplitude is considered by the Feynman coefficient α, which, in our case, is related to the frictional parameters by linear equations of the form (4.34). Considering (4.21), (4.22) and (4.34), the equation for the tribosystem wave function is transformed as follows:

$$\psi_R = 2\left\{\psi_{11} + \left[\psi_{11} - 2af\psi_{11} + (1 - 3af)\frac{\delta}{\omega}\dot{\psi_{11}}\right]\right\} \tag{4.35}$$

Equation (4.35) determines the functional relationship between the quantum state of the manifested tribosystem, characterized by ψ_R, and the frictional macro-parameters, such as friction coefficient. The friction coefficient f from (4.35), under the laws of wave physics and oscillations theory, is related with relaxation time τ and the lasing properties of the oscillators by a relation $\tau^{-1} = fm^{-1}$ [67]. Since the oscillators for the tribosystem considered are physical associates in complex interaction, consisting of many components, they can be considered as virtually isolated quasiparticles, the inertial characteristics of which are determined by a quantity called *effective mass* : $m^{-1} = \hbar^{-1}\frac{\partial \theta}{\partial \kappa}$, where θ is a collective velocity of the quasiparticle, κ is its wave number [70]. Under the calculations [71], the mean value m for a tribosystem has an order of 10^{-26} kg. Then the friction coefficient for the quantum system describing frictional interaction, in particular, of steel and polytetrafluorethylene near the vertex *1* of the Feynman diagram equals 0.2. This is close enough to the experimental results registered in run-in mode. The friction coefficient for this mode is obtained by additional multiplication of f by 0.2, which gives 0.04, with experimental data 0.052 and gives evidence of a good fit between the theory and the experiment.

Thus, the quantum mechanism of the tribosystem evolution allows to represent this process as changes that occur with a probability wave, which between the vertices of the Feynman diagram (Fig. 4.11) acquires a phase shift δ. In the very vertices, where, by definition, the interactions occur between material particles forming a physical system, the wave function amplitude decreases, as well as the probability of the following events.

The representation of the external force acting on the manifested tribosystem as the consequences of the quantum mechanisms described by perturbation theory, as well as the results we obtained before, adequately describe, using this algorithm, an interesting phenomenon for friction microdynamics and called 'field effect'.

Definition 4.8. *Field effect* is a change in the friction coefficient and the wear rate under the action of electromagnetic, thermal, acoustic and other physical fields [3]. This effect is especially important when studying the superexcited state of the friction surfaces substance, kind of triboplasma [59].

The physical consideration of the impact of physical fields on the kinetics of tribophysical processes, based on applying continuum electrodynamics, is a complicated task from the mathematical point of view [72]. Therefore, most works dedicated to this problem are limited to qualitative description of the physical aspect of this phenomenon. At the same time, the formalism of quantum mechanics, where all the information about a physical system is hidden in the wave function ψ, at least mathematically allows to construct a simple and physically correct algorithm that describes the field action on a tribosystem.

In accordance with the quantum calculation algorithm [37], one should assume that the wave function ψ describing the tribosystem behavior, as stated above, has a form of plane wave $\psi_0 \exp(i\omega t)$. Then the external force action, under (4.18), (4.19), results in appearance of a phase δ in the exponent:

$$\psi' = \psi_0 e^{i(\omega t+\delta)} = \psi_0[\cos(\omega t + \delta) + i\sin(\omega t + \delta)] \tag{4.36}$$

Using trigonometry, (4.36) is transformed:

$$\begin{aligned} \psi' &= \psi_0[\cos\delta(\cos\omega t + i\sin\omega t) - \sin\delta(\sin\omega t - i\cos\omega t)] \\ &= \psi\cos\delta + \omega^{-1}\sin\delta\cdot\dot{\psi} \end{aligned} \tag{4.37}$$

where ψ, ψ' is the wave function before and after external field action on the system; $\dot{\psi}$ is the time derivative of the initial wave function.

Using Euler's formula from complex analysis and based on equations (4.36) and (4.37), we may write:

$$\begin{gathered} \psi\cos\delta + i\psi\sin\delta = \psi\cos\delta + \omega^{-1}\sin\delta\cdot\dot{\psi} \\ \text{or} \\ i\psi = \omega^{-1}\dot{\psi} \end{gathered} \tag{4.38}$$

The differential equation (4.38) implies that the wave function ψ retains its initial form of a plane wave after the external field action on the physical system. Since the plane wave nature of ψ-function is not changed under external forces, the act of this action may be conventionally regarded as a quantum analog for perfectly elastic impact, under which there is no change in the internal structure of the physical system. Therefore, the external field action may be represented as a particle (quasiparticle) elastically colliding with the 'shell' of the quantum tribosystem [45]. It may be considered established that the probability wave distributed in space-time continuum does not change its form under the impact of internal and external forces. This conclusion itself contains an important assumption that was not talked over when considering dynamic interaction. It was implicitly assumed that force fields do not affect the metric of Minkowski space, which is not obvious from the viewpoint of relativistic physics. By definition, the wave function depends on continuum metric, but the metric itself may change under force action. Therefore, it is necessary to study the impact of relativistic factors on the wave function of the manifested tribosystem.

4.4 Relativistic Corrections to the Tribosystem Wave Function

Let us see what happens to the wave function under the impact of forces that cause relativistic effects. The formal manifestation of relativistic effects and their impact on the wave function parameters are considered by multiplying the initial wave function ψ by the phase factor $\psi_p = \exp\left[\frac{i}{\hbar}(W_p't' - p'x')\right]$, where $W' = W\beta^{-1}$; $t' = t\beta^{-1}$; $p = \frac{W'\theta x'}{c^2}$; $x' = \frac{(x-\theta t)}{\beta c^2}$; $\beta = \sqrt{1 - \frac{\theta^2}{c^2}}$; θ is the velocity of the physical body; c is the speed of light; W and t are the energy and the time of the body in rest, respectively; W' and t' are the energy and the time of the body moving with the velocity θ, respectively; p' is the relativistic momentum; x' is the coordinate of the body that moves with the velocity θ in the reference frame; x is the coordinate of the body in the stationary frame of reference.

Then the impact of relativistic forces on the physical system may be represented as [37, 48]:

$$\psi' = \psi \cdot \psi_p = \psi_0 e^{i\omega t}\psi_p \tag{4.39}$$

For convenience, we rewrite the plane wave describing the initial state of the physical system in terms of energy W, the relation (4.39) taking the form:

$$\psi' = \psi_0 e^{\frac{i}{\hbar}(Wt + W't' - p'x')} \tag{4.40}$$

Substitute the values of the physical quantities described earlier with the exponent of (4.40) and, with the natural for relativistic mechanisms condition $\theta < c$ in mind, we get:

$$\frac{i}{\hbar}(Wt + W't' - p'x') = \frac{i}{\hbar}Wt(0.5\beta - \beta^{-1}) \tag{4.41}$$

Since $\beta < 1$, $(0.5\beta - \beta^{-1}) < 0$, the exponent changes its sign:

$$\frac{i}{\hbar}(Wt + W't' - p'x') = -\frac{i}{\hbar}Wt(\beta^{-1} - 0.5\beta) \approx -\frac{i}{\hbar}\frac{Wt}{\beta} \tag{4.42}$$

Then, with (4.42) in mind, the wave function (4.40) takes the form:

$$\psi' = \psi_0 e^{-\frac{i}{\hbar}\frac{Wt}{\beta}} \tag{4.43}$$

Even a superficial analysis of the relation (4.43) shows that the impact of 'relativistic forces' on a system significantly changes the appearance of the wave function that describes the state and development of the physical object (a tribosystem in our case). These changes formally change the exponent sign and the appearance of the relativistic parameter β in the denominator. However, behind the formal aspect there are very complex physical regularities we need to analyze deeply.

The change of the exponent sign in (4.43) may be explained by two reasons: the appearance of a negative mass and the change in temporality of physical processes. Both explanations sound very unusual for one not familiar with the canons of modern physics, though they completely agree with its fundamental laws [26]. V.S. Barashenkov wrote: "...what is not prohibited by physical laws, may be a reality" [26]. As shown in [59, 66], the frictional interaction has a complex multiparticular nature, for the theoretical description of which we may use the ideas about quasiparticles introduced by the Soviet physicist, Nobel Prize-winner, L.D. Landau.

Definition 4.9. *Quasiparticle* is a fundamental concept in quantum theory, the introduction of which radically simplifies the physical picture of the world and the methods to describe a wide range of processes in systems of many particles with strong interaction. A quasiparticle is a special long-living multiparticular complex, which, unlike ordinary particles that form a system, weakly interacts with its environment. Therefore, a quasiparticle is in a special quantum state with its own wave function, energy, momentum, spin, etc. A quasiparticle, being a collective formation, is continuously renewing in motion composition that remains fixed only in extreme cases [73].

The fact that quasiparticles have a spin suggests that quasiparticles should obey the laws of quantum statistics, which divide them into two big classes – bosons and fermions. Fermions are particles or quasiparticles that have spin 1/2, are described by asymmetric wave functions and obey the Pauli exclusion principle, according to which, in a quantum system there cannot be more than one fermion [31]. That is, between fermions there are distinctive forces of 'quantum repulsion' that do not allow Fermi particles to form compact structures. These forces determined the specifics of atomic shells structure, hence all chemical properties of substances in the nature.

Unlike fermions, bosons not only have integer spin, their behavior is described by even wave functions. It allows several Bose particles to exist in the same quantum state. P. Dirac in his *Principles of Quantum Mechanics* wrote: "...the probability of a transition in which a boson is emitted into state x is proportional to the number of bosons originally in state x plus one... the probability of a transition in which a boson is absorbed from state x is proportional to the number of bosons originally in state x" [44]. Unlike Fermi particles, bosons are able to form compact structures. Such structures are called Bose condensate.

Definition 4.10. *Bose condensation* is a quantum phenomenon consisting of the fact that in a system with a big number of bosons, at temperatures below the degeneration temperature, a finite fraction of the quantum system particles is in a state with zero momentum [52].

The friction surface substance in a superexcited state may be regarded as a set of weakly interacting quasiparticles, which, under Def. 4.9 and 4.10, can have both integer and 1/2 spin and therefore form two groups of quasiparticles: tribobosons and tribofermions, obeying various laws of quantum statistics [31, 44]. Tribobosons should form compact systems, the Bose condensate, and tribofermions, based on uncertainty and exclusion principles, form structures similar to electron shells of atoms and molecules [37].

The formation process of a bunch of Bose condensates is regarded in detail in [74], which shows that united tribobosons form a bosonic sphere whose dimensions are determined by both quantum particles dimensions and quantum laws, in particular, the uncertainty principle. Tribobosons, with a spin greater than one, take part in the formation of the bosonic sphere, the size of the sphere, because of relatively small number of Bose particles ($\approx 10^{10}$) and does not exceed one millionth of the frictional interaction space volume. Despite its small dimensions, the Bose condensate has a huge density, about 10^{24} m^{-3}, which, under theory of relativity, allows to refer it to the factors that significantly impact the metric of Minkowski space-time. In fact, a 'drop' of the Bose condensate might be considered similar to a 'black hole' [53].

Definition 4.11. *Black hole* is an object in space that has an enormous gravitational field that holds absolutely all bodies, even light that tends to abandon it. A black hole is formed as a result of gravitational collapse, when the compression of an object reduces its original dimensions up to the so-called gravitational radius, or the Schwarzschild radius [55]:

$$R_G = 2\frac{GM}{c^2} \tag{4.44}$$

where G is the gravitational constant; M is the body mass; c is the speed of light.

When the gravitational radius is reached, the gravitational instability begins, which leads to a catastrophic contraction up to the singularity and related process of forming infinitely large density of matter and energy. The gravitational collapse leads to full information loss about the original object, all the prehistory of which disappears inside the Schwarzschild radius, which, to denote the act of information loss, is called an 'event horizon'. The formation of a black hole results in time dilation as the object approaches the Schwarzschild sphere and in its whole stop directly on the gravitational radius, inside of which space and time change places [23].

The answer to the question if an enormous density of the Bose condensate transforms the latter into a black hole of macroscopic dimensions is linked to the question: is it big enough so that the phenomenon of gravitational instability could arise, the theory of which has been formed mostly owing to the works of the English physicist and astronomer James Jeans [53].

Jeans, based on Newtonian gravitation theory and the equations of hydrodynamics, studied small density perturbations, the flow velocities and the gravitational potential in a homogeneous medium. He defined them as the waves determined by a value called Jeans wavelength λ_J. The λ_J quantity characterizes the minimum perturbation scale, starting from which the matter elastic forces are unable to resist the gravitational forces. As a result, the gravitational instability begins followed by uncontrolled sphere contraction and finally leading to the singularity surrounded by the event horizon and having space-time distortions. Under Jeans theory, the wavelength λ_J is calculated by the formula:

$$\lambda_J = c_s \sqrt{\frac{\pi}{G\rho}} \tag{4.45}$$

where c_s is the speed of sound; G is the gravitational constant; ρ is the medium density.

The speed of sound is defined by the relation $c_s = \dfrac{\theta}{\beta}$, where θ is the dislocation velocity; β is the relativistic correction calculated by tribo-activated Doppler shift (approximate order $\beta = 10^{-3}$). With this in mind, (4.45) may be simplified as:

$$\lambda_J \sim \frac{0.22 \cdot 10^5}{\sqrt{\rho}} \tag{4.46}$$

Since the Bose condensate density is much greater than 1, $\lambda_J < 0.22 \cdot 10^5$, then the Jeans mass M_J corresponding to the wavelength λ_J, equal to $\rho\lambda_J^3$ has an order:

$$M_J = \frac{10^{13}}{\sqrt{\rho}} \tag{4.47}$$

That is, the matter mass, with which the Jeans sphere begins its uncontrolled gravitational contraction and, under frictional interaction, turns into a black hole must not exceed 10^{13} kg. If a sphere of radius λ_J out of the Bose condensate matter is made, the mass of this formation will equal 10^{39} kg, which will obviously assure the gravitational collapse. Hence, after the decoherence and the birth of the manifested tribosystem, a microscopic object appears in its space that is able to significantly distort the metric of Minkowski continuum, creating temporal perturbations inside the friction area.

It is important that in the mid of the last century, the Soviet theoretical physicists, Y.B. Zel'dovich and I.D. Novikov considered in detail the possibility of forming a black hole out of individual particles of dust cloud, which stuck together forming a body of related mass only under the gravitational forces. This was quite sufficient for the beginning of the gravitational contraction process and the formation of a structure of the size of the Schwarzschild sphere [55]. The physical system considered by Zel'dovich and Novikov is mostly identical to the bunch formed by the tribobosons flow. Therefore, their conclusions about the arising of an object of 'black hole' kind in dust particles condensation are true in our case of appearance of a tribocollapsar out of the Bose condensate [74].

The bosonic tribocollapsar significantly changes the metric of Minkowski space immediately close to the point of its localization. These changes are also registered at far distances separating the instrument from the friction unit where that quantum-relativistic object arose. The distinctive changes in space-time may be discovered with a set of simple experiments described by Prof. N.A. Kozyryev [29]. These include direct weighing, study of physical pendulum oscillations, as well as a set of experiments on changes in the system entropy: heat elimination and dissolution of various substances. These experiments were added by a study of the changes in frequency characteristics of a quartz oscillator and conducted in [61]. The results of this experimental work allowed a conclusion about the existence of a certain force (quasi-gravitational) anomaly inside the friction unit, which might be a tribocollapsar.

The occurrence of such an object as a 'micro'-black hole results in separation of Minkowski continuum into two subspaces with different properties and, above all, their own local time. The existence of physical systems in separate parts of which time flows differently does not contradict the laws of physics. For example, the eminent Russian physician, Nobel Prize-winner Andrey D. Sakharov developed a theory of a world with infinite number of time dimensions: "...many things witness that our universe is really multidimensional not only in space, but also in time; those additional coordinates are still simply hidden from our instruments" [75].

Digression 4.3. The famous English physicist, Arthur Eddington supposed that time flow is determined by the modern stage of the universe development. He assumed that the evolution of the universe is cyclic and at a definite moment. When the contraction replaces the expansion, a change will be in the time vector direction. In Eddington's model, space and time are born immediately at the 'big bang' moment out of the singularity point [29].

Eddington's views on the birth of time and space out of singularity might be applicable to the physical model of the evolution of a tribosystem born out of non-local information state in the decoherence process and manifested in Minkowski world as a 'compound' space-time object consisting of two subsystems: the strongly non-Euclidean curved continuum near the Schwarzschild radius of the tribocollapsar and the 'relatively Euclidean' Minkowski space where the tribofermions are distributed as in (Fig. 4.12) [74].

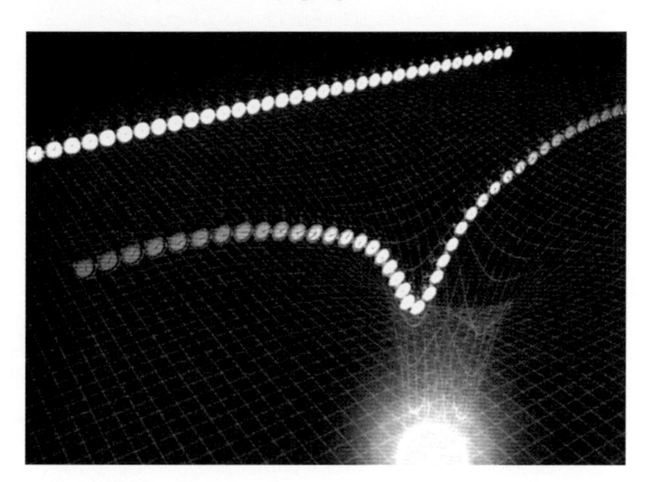

Figure 4.12: Heterogeneous structure of space-time continuum illustrating the co-existence of two subsystems: below it can be seen how space-time curves under the impact of energy of the born tribocollapsar

Chapter 5

Quantum States of Manifested Tribosystem

Any quantum particle (or a system presented as a quasiparticle) at rest has a probability amplitude that characterizes the possibility to find this system in any point of phase multidimensional space. Since this particle (quasiparticle) is at rest by definition, the amplitude of this probability should not depend on the particle localization. The concept 'at rest' for a quantum system, due to the uncertainty principle, is conventional; hence it is in 'quotes'. The independence of the probability amplitude value from the coordinate of the particle at rest means that the probability to find this object at any point of space is the same, but at the same time, the phase of this amplitude still may change from point to point.

Although the probability amplitude for a system at rest is equal at all points of space, there is as important parameter as time in the relations that determine it. And that would do, if there was no laws of relativity theory, under which there is a dependence of time flow on the velocity of the object motion and the space geometry [55]. Since even for a particle at rest there may always be a system with respect to which our physical system is moving, from the most general considerations, the distribution of its probability amplitude changes. Therefore, it is necessary to not only know the state of the physical system at a given moment, but also calculate the probability (or 'chances', in terms of Feynman) that its 'path' (or history) would pass through definite points of space-time. For that it is necessary to establish the physical relations (laws) that determine those regularities, which is a quantum similarity of dynamic equations in classical mechanics.

The basic equations of the laws of 'quantum dynamics' have two universally recognized notation forms: the matrix form starting from works by W. Heisenberg and M. Born, and the wave equations from E. Schrödinger's representations [28]. The relations between physical quantities necessary to start considering the dynamic regularities of a tribosystem development are cited in Chapters 2 and 3 as postulates, theorems and definitions, which determine the relationship between quantum parameters.

5.1 Secular Equation of Manifested Tribosystem

Information about the parameters cross-effect requires establishment of 'special dynamic relations' between them. For that consider a tribosystem as a three-level quantum object in accordance with the features of the friction stages. Based on that, a tribosystem may be one of the three basic states: $\left|1\right\rangle, \left|2\right\rangle$, or $\left|3\right\rangle$. Hence, the true state of the tribosystem $\left|\psi\right\rangle$ may be represented by defining the corresponding amplitude $\left\langle 1\middle|\psi\right\rangle = c_1$ of its being in state $\left|1\right\rangle$, $\left\langle 2\middle|\psi\right\rangle = c_2$ of being in state $\left|2\right\rangle$, and respectively, $\left\langle 3\middle|\psi\right\rangle = c_3$ of being in state $\left|3\right\rangle$.

From the superposition principle:

$$\left|\psi\right\rangle = \left|1\right\rangle\left\langle 1\middle|c_1\right\rangle + \left|2\right\rangle\left\langle 2\middle|c_2\right\rangle + \left|3\right\rangle\left\langle 3\middle|c_3\right\rangle = c_1 + c_2 + c_3 \tag{5.1}$$

For each of these states we may write its own equation (the Schrödinger equation) [24]:

$$\begin{aligned} i\hbar\frac{dc_1}{dt} &= H_{11}c_1 + H_{12}c_2 + H_{13}c_3 \\ i\hbar\frac{dc_2}{dt} &= H_{21}c_1 + H_{22}c_2 + H_{23}c_3 \\ i\hbar\frac{dc_3}{dt} &= H_{31}c_1 + H_{32}c_2 + H_{33}c_3 \end{aligned} \tag{5.2}$$

where H_{ii} and H_{ij} are the Hamiltonians; i is the imaginary unit.

Let our quantum tribosystem be initially in the basic state $\left|1\right\rangle$. In a certain amount of time, it transits to state $\left|2\right\rangle$, then to state $\left|3\right\rangle$. To define the probability of such transitions, we should solve the system of equations (5.2). As shown in [24], with respect to the Hamiltonians, we may admit the equality $H_{ij} = H_{ji}$ and prohibit the system to make quantum transitions through the basis—the system may transit from the first state to the second or vice versa; but it may not transit from the first state to the third and vice versa, bypassing the second, etc. Let us define:

$$\begin{gathered} H_{11} = W_1, H_{22} = W_2, H_{33} = W_3 \\ H_{12} = H_{21} = H_{23} = H_{32} = A \\ H_{13} = H_{31} = 0 \end{gathered} \tag{5.3}$$

where W_i are basic states energies; A is the transition energy between the bases.

Summing up (5.2) under (5.3), we get:

$$i\hbar\frac{d}{dt}(c_1 + c_2 + c_3) = i\hbar\frac{d}{dt}c_\Sigma = c_1(W_1 + A) + c_2(W_2 + 2A) + c_3(W_3 + A) \tag{5.4}$$

We represent (5.4) as the Schrödinger equation:

$$i\hbar\frac{dc_\Sigma}{dt} = W_\Sigma c_\Sigma = W_\Sigma(c_1 + c_2 + c_3) \tag{5.5}$$

Then, to keep the canonical form of the quantum equation (5.5), it is necessary that $W_1 + A = W_2 + 2A = W_3 + A$. At the same time:

$$\begin{gathered} W_1 = W_3 \\ W_1 = W_3 = W_2 + A \end{gathered} \tag{5.6}$$

The wave functions from (5.2), according to data from Section 4.2, have the form of plane waves:

$$c_i = a_i e^{-\frac{i}{\hbar}W_t} \tag{5.7}$$

Substituting (5.7) in the system (5.2), we get a set of equations:

$$\begin{gathered} Wa_1 = W_1a_1 + Aa_2 + 0 \\ Wa_2 = Aa_1 + W_2a_2 + Aa_3 \\ Wa_3 = 0 + Aa_2 + W_3a_3 \end{gathered} \tag{5.8}$$

The presented equations may be transformed as follows:

$$
\begin{aligned}
(W - W_1)a_1 - Aa_2 - 0 = 0 \\
-Aa_1 + (W - W_2)a_2 - Aa_3 = 0 \\
0 - Aa_2 + (W - W_3)a_3 = 0
\end{aligned}
\tag{5.9}
$$

To find the energy value for this system, mark its determinant Δ_W and set it equal to zero.

$$
\Delta_W = \begin{vmatrix} W - W_1 & -A & 0 \\ -A & W - W_2 & -A \\ 0 & -A & W - W_3 \end{vmatrix} = 0 \tag{5.10}
$$

The solution of (5.10) for W results in:

$$
\begin{aligned}
W^{(1,2)} = \frac{W_2 + W_3}{2} \pm \frac{W_3 - W_2}{2} \\
W^{(1)} = W_1 = W_3 \\
W^{(2)} = W_2
\end{aligned}
\tag{5.11}
$$

This result has quite an obvious physical meaning: the Hamiltonian W in the process of the tribosystem interaction where the quantum tribometer may have a set of its own energy values relative to the states in the Feynman diagram (Fig. 4.11).

For a complete description of a quantum system, we should find the values of the coefficients a_i in the system of secular equations (appendix to Theorem 3.11), which, according to [48], are the probability amplitudes of the quantum system transitions from state W to the states defined by own values of the Hamiltonians W_1, W_2, W_3, which may be considered observable, i.e. the quantities obtained with some degree of probability at direct physical measurements. From the probability interpretation of the coefficients a_i, we may assume that the following equality is met:

$$
a_1^2 + a_2^2 + a_3^2 = 1 \tag{5.12}
$$

From the symmetry based on the solution results (5.11) of the determinant (5.10) indicating the equality of the states W_1 and W_3, we may assume that $a_1 = a_3$. Then the normalization (5.12) will be transformed as:

$$
a_1 = \frac{1}{\sqrt{2}}(1 - a_2^2)^{\frac{1}{2}} \tag{5.13}
$$

In accordance with (3.72), the coefficient $\frac{1}{\sqrt{2}}$ that appears in (5.13) may be treated as a wave vector rotation by 45° at the transition from one phase subspace of the tribosystem to another, which in fact occurs, under the results obtained in Section 3.4.

To find the values of the coefficients a_i that determine the histories (in Feynman's sense) of the tribosystem components and, accordingly, the features of the phase subspace metric where the manifested tribosystem 'materialized' because of the decoherence, consider again the system of equations (5.9). That system, from the viewpoint of linear algebra, is a homogeneous system of three equations of the first degree with three unknowns [51]. We rewrite (5.9):

$$\begin{aligned} a_1(W_1 - W) + a_2 A = 0 \\ a_1 A + a_2(W_2 - W) + a_3 A = 0 \\ a_2 A + a_3(W_3 - W) = 0 \end{aligned} \tag{5.14}$$

In such systems of equations, the free terms are equal to zero. It is obvious that the system (5.14) always has a trivial solution: $a_1 = a_2 = a_3 = 0$, called null solution. If the determinant of the homogeneous system is non-zero, this solution is called unique [51]. But if it equals zero, the homogeneous system (5.14) has an infinite set of non-zero solutions, because in that case either some of the equations are a consequence of two others or some two equations are a consequence of the third one.

Let the two first equations in the system (5.14) have infinitely many joint non-zero solutions defined by the formulae [51]:

$$\begin{aligned} a_1 &= \begin{vmatrix} A & 0 \\ W_2 - W & A \end{vmatrix} t \\ a_2 &= -\begin{vmatrix} W_1 - W & 0 \\ A & A \end{vmatrix} t \\ a_3 &= \begin{vmatrix} W_1 - W & A \\ A & W_2 - W \end{vmatrix} t \end{aligned} \tag{5.15}$$

with any value of the parameter t.

We get the solution of (5.15) thus:

$$\begin{aligned} a_1 &= tA^2 \\ a_2 &= -t(W_1 - W)A \\ a_3 &= t\left[(W_1 - W)(W_2 - W) - A^2\right] \end{aligned} \tag{5.16}$$

Then from the solutions of (5.11) for $W = w^{(1)}$, the probability amplitudes a_i from (5.16) are $a_1 = tA^2, a_3 = -tA^2, a_2 = 0$. From (5.12), $a_1 = |-a_3| = \frac{1}{\sqrt{2}}$. For the Hamiltonian $W = W^{(2)}$ all the amplitudes under (5.6) are equal modulo $+tA^2$. From their normalization (5.12), their numeric values are $a_1 = -a_2 = -a_3 = 1/\sqrt{3}$.

Hence, the manifested tribosystem may have the structure described by both the first set of amplitudes and the more complex second configuration relative to the Hamiltonian $W = W^{(2)}$. The logic of quantum physics guided by the 'immortal' Schrödinger cat affirms that the states with Hamiltonians $W^{(1)}$ and $W^{(2)}$ are not alternative; they co-exist simultaneously, being equal terms of quantum superposition. But they are localized in different subspaces of the Hilbert space of the quantum tribosystem, manifesting themselves in interaction of the latter with a quantum tribometer (Fig. 5.1).

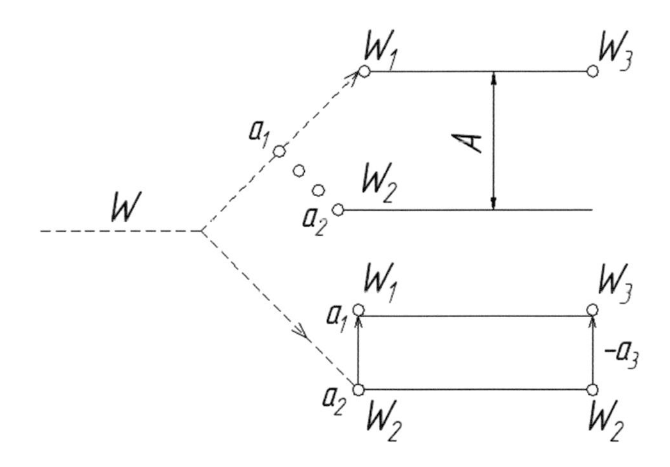

Figure 5.1: Quantum diagram of the energy structure of a tribosystem (symbols from Feynman diagrams are used: broken lines denote virtual states and transitions; points denote the forbidden transition)

Special attention is required in explaining the negative value of the amplitudes a_i. In his time, Dirac, having seen negative energy values, defined them as a manifestation of the physical vacuum as described in Section 3.4. The matter of the physical vacuum is in a virtual state [53], out of which the reverse interactions quanta are born together with the antiparticles that make it possible to move backwards in time or make reverse transitions in space, against the rules that obey the energy conservation law. Anyway, the negative amplitude values prove the presence of virtual areas in the evolution of tribosystems postulated in [37, 45] to ground the features of entropic development of friction units.

5.2 Quantum Representations of Mechanical Parameters of a Tribosystem

Section 5.1 establishes the dependencies between the energies of various states of a manifested tribosystem, which, in turn, is a product of reduction of a quantum non-local 'properly tribosystem'. The Hamiltonian of the manifested tribosystem whose eigenvalues are the energies from the secular equation (5.2), under Lemma 2.1 and Theorem 3.13, should commute with the original Hamiltonian of properly tribosystem:

$$\lfloor H_i H_j \rfloor = H_i H_j - H_j H_i = 0 \tag{5.17}$$

where H_i, H_j are the commutators of different kinds of tribosystems.

Let us make an assumption about the forms of the Hamiltonians H_i and H_j. Since the wave functions describing the state of tribosystems are plane waves, similar to those free particles are described by, as a first approximation, it is logical to assume that $H_i = \dfrac{P_i^2}{2m}$. The latter in one-dimensional approximation is

$$H_i = \frac{\hbar^2}{2m}\frac{\partial^2}{\partial x^2} \tag{5.18}$$

If the other Hamiltonian H_j is a linear function of mechanical work A (including friction forces), then

$$H_j = \alpha A(x) \tag{5.19}$$

where $A(x)$ is the mechanical work operator; α is a real-valued coefficient.

Substitute (5.18) and (5.19) in (5.17):

$$\frac{\partial^2 A(x)}{\partial x^2} = A(x)\frac{\partial^2}{\partial x^2} \tag{5.20}$$

Multiply the left and right parts by the wave function $\psi(x)$:

$$\begin{aligned} \psi(x)\frac{\partial^2 A(x)}{\partial x^2} &= A(x)\frac{\partial^2 \psi(x)}{\partial x^2} \\ A^{-1}(x)\frac{\partial^2 A(x)}{\partial x^2} &= \psi^{-1}(x)\frac{\partial^2 \psi(x)}{\partial x^2} \end{aligned} \tag{5.21}$$

We introduce a new variable $z(x) = \psi(x)/A(x)$ and, with that quantity in mind, rewrite (5.21):

$$z''(x) = z(x) \tag{5.22}$$

The condition (5.22) may be met only if the function $z(x)$ equals the exponent e^x. Hence, mechanical work is represented as a product:

$$A(x) = \psi(x)e^{-x} \tag{5.23}$$

Taking into consideration that the wave function $\psi(x)$ is itself a plane wave, i.e. is written as the exponent $\psi_0 e^{-i\kappa x}$, (5.23) is rewritten as:

$$A(x) = \psi_0 e^{-(1+i\kappa)x} \tag{5.24}$$

where κ is the wave vector.

The quantity $A(x)$ is expressed in tribological parameters by the well-known relation [66]:

$$A(x) = \beta' \cdot f \cdot N \cdot x_{mp} \tag{5.25}$$

where β' is a numeric coefficient considering the dissipative processes in friction, it has a numeric value close to 10^{-2} [16]; N is the normal load; f is the friction coefficient; x_f is the friction path.

If we combine the parameters N and x_f as the macro-characteristics of the friction conditions and denote it as a_e^{-1}, from (5.24) and (5.25) the relation for the friction coefficient is derived:

$$\begin{aligned} f &= f_0 \exp\big[-(1+i\kappa)x\big] \\ f_0 &= 10^2 a_e \psi_0 \end{aligned} \tag{5.26}$$

The special case of (5.26) is harmonic oscillations, which, according to the data from [37], well describe the well-known tribology phenomenon of ‘frictional oscillations’. The fact that the wave function describing the tribosystem state retains its plane wave form allows using the free particles approximation. The dependence of the system energy on the wave vector κ of de Broglie wave has a quadratic form [37]:

$$W = \frac{\hbar^2 \kappa^2}{2m} \tag{5.27}$$

where $\hbar$ is the Planck's constant; κ is the wave vector; m in general case is the effective mass of the physical system [70].

Knowing that Minkowski space-time is homogeneous and isotropic and $m^{-1} = \frac{\partial^2 W}{\partial \kappa^2}$, the expression (5.27) may be rewritten as:

$$\begin{gathered} W = \frac{\kappa^2}{2} \frac{d^2 W}{d\kappa^2} \\ \frac{d^2 W}{d\kappa^2} - 2\kappa^{-1} \frac{W}{\kappa} = 0 \end{gathered} \tag{5.28}$$

The differential equation (5.28) describes harmonic oscillations [67]. Hence, its solution is as follows:

$$\begin{gathered} \frac{W}{\kappa} = \frac{W_0}{\kappa_0} e^{-\sqrt{\frac{2}{\kappa}}} \\ W = W_0 \frac{\kappa}{\kappa_0} e^{-\sqrt{\frac{2}{\kappa}}} \end{gathered} \tag{5.29}$$

Equation (5.29) for dispersion relations [59] characterizes the changes in energy states of the matter inside the frictional contact space. The wave vectors relation in (5.29) may be substituted by relation of the relevant oscillation frequencies ω/ω_0. The difference between ω_0 and ω is caused by the impact of dissipative forces shown in Section 4.3. In oscillation theory it is described by a value called generalized coefficient of friction $\gamma = \gamma(\eta, f)$, which depends on internal and external friction (4.34)[67]:

$$\frac{\omega}{\omega_0} = \frac{\sqrt{\omega_0^2 - \gamma^2 m^{-2}}}{\omega_0} = 1 - \frac{1}{2} \frac{\gamma^2}{m^2 \omega_0^2} = \frac{1}{2}\left(1 - \frac{\omega^2}{\omega_0^2}\right) \tag{5.30}$$

Based on the representation of the wave-corpuscle dualism by de Broglie [53], any particle may be regarded as a wave packet that moves with group velocity θ [50].

Note 5.1. The representation of a particle as a wave packet has a significant disadvantage. Numerous theoretical models show that the wave packet is not stable and tend to 'spread' across the space, which contradicts the locality peculiar to all particles. Therefore, considering individual corpuscles as wave packets reflects only the aspect of quantum wave matter properties that all particles should have wave properties. Hence, particles representation as wave packets may be considered a first approximation.

The group velocity for the frequency ω_0 from formula (5.30) equals the speed of light in free space c. Then the latter expression may with a certain approximation be considered relativistic correction $\beta = \sqrt{1 - \frac{\theta^2}{c^2}}$. This representation allows to rewrite the expression for the proper energy values as:

$$W = \beta W_0 e^{-\sqrt{\frac{2c}{\omega}}} \tag{5.31}$$

(5.31) indicates that the relativistic effects [59, 45] that manifest themselves at microlevels of tribophysical processes make a significant impact on the parameters characterizing the states of the manifested tribosystem. If we assume that the physical states that describe the behavior of ordinary particles are valid for their quasi-analogs, we may write the expression for the effective mass as:

$$m = \frac{\hbar\omega}{c^2} = \frac{p}{c} \tag{5.32}$$

where p is the quasiparticle momentum as a characteristic of frictional interaction [37, 45].

In (5.32) the relation m/p is a constant value equal to c^{-1}. If we assume that in tribomutation processes [76] the values of momentum and energy for quasiparticles remain the same (the latter follows from quantum mechanics equations on the one hand and the conservatism of friction forces, on the other), yet another integral of motion characterizing the behavior of the manifested tribosystem may be effective mass m (5.32). All changes in frictional characteristics of the friction unit are determined by the parameter $\gamma = m\tau^{-1} = m\omega$ [67], owing to which, we may rewrite (5.30) as:

$$\gamma = m\sqrt{\omega_0^2 - \frac{\gamma^2 W^2}{\hbar^4 \kappa^4}} = m\frac{\sqrt{\hbar^4 \omega_0^2 \kappa^4 - \gamma^2 W^2}}{\hbar^2 \kappa^2} \tag{5.33}$$

Considering ω_0 and ω close enough, transform (5.30) to $W\sqrt{\hbar^2\kappa^4 - \gamma^2}$, where $W = \hbar\omega_0$. Then, from (5.27), the relation (5.33) takes the form:

$$\gamma = \frac{\sqrt{p^2\kappa^2 - \gamma^2}}{2} \tag{5.34}$$

$$\gamma = 0.5p\kappa = 2m\omega \tag{5.35}$$

The formula (5.35) reveals the physical meaning of the parameter γ and the micromechanisms of friction processes. From this relation, the generalized friction coefficient is a consequence of an impulse action on the tribosystem counteracting the translational motion of a particle of quasiparticle. The nature of the impulse as a physical quantity, its relations with the parameters characterizing the discrete matter structure allow consideration of (5.35) as one of the proofs of quantum mechanism of frictional interaction of solid surfaces [37].

Definition 5.1. *Quantum mechanism of frictional interaction* is based on the assumption that the friction force may be represented as an exchange of special quasiparticles, tribons, the quanta of frictional interaction, between the surfaces. Theoretical publications dedicated to tribons generation in solid surfaces contact show that they are born in pairs – one tribon a surface [37]. The reflection of this paired nature of tribons and the confirmation of this statement is the coefficient 2 in (5.35).

The tribonic mechanism of frictional interaction explains the additivity of friction forces [8], since the macro-manifestations of those forces may be explained by combined action of multiple tribon pairs making a joint momentum. Then, according to data from [67], the macro-coefficient of internal friction η may be written as:

$$\eta = m\sqrt{\omega^2 - \frac{\eta^2}{N^2 m}} = 0.5p\kappa\sqrt{N} = \gamma\sqrt{N} \tag{5.36}$$

The representation of friction coefficient γ in impulse form (5.35) in the analysis of this physical quantity allows use of the quantum mechanic motion equation (3.74) in the form:

$$\frac{d\gamma}{dt} \sim [pH] \tag{5.37}$$

In accordance with the matrix form of the commutator $[pH]$ (3.77), where the operators p and H commute with each other, the expression (5.37) vanishes:

$$\frac{d\gamma}{dt} = 0 \tag{5.38}$$

The equality of the time derivative of the parameter γ to zero shows its independence from time and, accordingly, the possibility to regard that physical quantity as integral of motion. The known time macrocharacteristics of friction $\eta(t)$ and $f(t)$, from (5.36), should depend on the parameter N – the number of tribons generated. The obtained result allows to assume that an act of frictional interaction in quantum mechanics is worth considering with the Schrödinger picture [49, 50].

Digression 5.1. There are two approaches to the notation of the motion equation in quantum mechanics. They are called Schrödinger and Heisenberg pictures. In the Schrödinger picture, the variable in time 'vector' components of wave functions $\psi(t)$ are defined in Hilbert space with a basis independent from time. In other words, any independent from time variable is described by a matrix on the basis of that space. Although the elements of that matrix do not depend on time, the chance to obtain definite results in measurements at a moment t in fact depends on t. Hence, the wave function is itself a state vector of the quantum system and depends on time. It is postulated that Minkowski space-time is continuous. This description of quantum systems is called Schrödinger picture, since E. Schrödinger in 1926 gave the original formulation of quantum mechanics [44]. The Schrödinger picture exists only in a system where canonical transformations are possible [44] and those transformations are defined up to the phase factor. For example, there are canonical coordinates q and p, then there exists a known relation $[qp] = i\hbar$ between them.

From the Heisenberg picture, the state vector ψ, dependent on time, is related to its original value $\psi(0)$ by unitary transformation: $\psi(t) = Q(t)\psi(0)$. That is, the dependence of a physical quantity on time is linked with the basis of Hilbert space. The components of the wave vector ψ are equal to $\psi(0)$, hence, they do not depend on time; that is, they seem to be defined in the space of Schrödinger picture. It is said that, in the Heisenberg picture, the state vector does not depend on time. The Heisenberg picture of quantum mechanics is the closest to classical ways of describing the process of motion.

The relations (5.35)–(5.38), supplemented by the conclusions from formula (4.33), lead us to the Schrödinger picture of the physical quantity $\eta(t) = \gamma\psi(t)$. The wave function ψ, from (4.33) is equal to $\sqrt{N}$. The number of particles N determines the time dependence $\eta(t)$.

The existence of the micro-coefficient of friction γ as an integral of motion turns us back to the initial works on friction and wear done by the great Italian artist, architect, engineer and scientist of the Renaissance, Leonardo da Vinci, who considered the friction coefficient a constant value 0.25 [3]. Obviously, this coincidence is accidental, but we assume that the micro-coefficient of friction γ may be called the 'Leonardo' coefficient.

5.3 Macrocharacteristics of a Manifested Tribosystem

In (5.35), friction at quantum levels of tribosystem organization may be regarded as the processes of tribon dispersions over atoms or tribons over tribons, which is well illustrated by the Feynman diagrams [45]. These processes change the absolute value of the wave vector ψ, which, as Feynman said, is interpreted as 'arrow shrinks' [54]. This phenomenon reduces the probability of tribosystem transition into a finite state and is illustrated by introducing the coefficient α in (4.18), the procedure that was then considered formal. In fact, this coefficient, present in quantum equations describing tribosystems microevolutions, serves as a numerical characteristic of the friction and wear processes. Therefore, a priori between the coefficient α and the traditional parameter that characterizes frictional processes, a linear relationship was established (4.34). The idea of micromechanisms of frictional interaction as acts of particles or quasiparticles dispersion requires theoretical supplements and comments.

Supplement. The calculation of particles dispersion is done by introducing the initial amplitude ψ to the expression for plane wave, describing the state of the new member of the quantum system that corresponds to the diverging spherical wave that describes the dispersion effect of the wave front as an arbitrary obstacle [68]:

$$\psi(x,t) = e^{i\kappa x}\lfloor 1 + x^{-1} f(\Theta) \rfloor \tag{5.39}$$

where Θ is the dispersion angle; $f(\Theta)$ is the dispersion function.

It is important that the wave function $\psi(x,t)$ of the form (5.39) being the solution for the Schrödinger equation that describes the behavior of free particles, finally allows use of the wave numbers from (5.34)–(5.35). By definition, the dispersion function $f(\Theta)$ is expressed by probability flow density [31, 68] and is determined by dispersion cross-section σ:

$$\dot{f}(\Theta) = \int \frac{\partial \psi^*}{\partial t} \psi dV + \int \psi^* \frac{\partial \psi}{\partial t} dV \tag{5.40}$$

$$|f(\Theta)|^2 = \frac{d\sigma}{d\Omega} = \frac{\sigma}{4\pi} \tag{5.41}$$

where Ω is the dispersion solid angle.

Based on the expressions presented, the wave function (5.38) has the following form:

$$\psi(x,t) = e^{i\kappa x}\left[1 + \frac{1}{2x}\sqrt{\frac{\sigma}{\pi}}\right] \tag{5.42}$$

The dispersion cross-section σ is determined by a whole set of external factors that impact the dispersion processes:

$$\frac{\sigma}{4\pi} = \frac{4}{\chi^4}\left(\frac{\alpha_e}{p\theta}\right)^2 \tag{5.43}$$

where χ is the deflection angle of the particle dispersed; α_e is the energy constant; p is the momentum; θ is the particle velocity.

The dispersion cross-section is related to the mean free path of the particle $\bar{x}$:

$$\begin{gathered} \bar{x} = \sqrt{\frac{\sigma}{\pi}} = \frac{\gamma}{\hbar\kappa} \\ \psi(\bar{x},t) = \frac{3}{2} e^{i\kappa\bar{x}} \end{gathered} \tag{5.44}$$

As often stated above, the manifested tribosystem evolves in Minkowski continuum, homogeneous and isotropic in all properties over all directions. There are three of them in the Euclidean space. That is why in (5.44) the proper component appears, the equation of the relation between the 'quantum coefficient of friction' γ and the mean free path $\bar{x}$ is written as:

$$\gamma = \frac{1}{3}\hbar\kappa\bar{x} \tag{5.45}$$

With (5.36) in mind, the coefficient of internal friction η for the frictional contact microsystem is written in the following way:

$$\eta = \frac{1}{3}\hbar\kappa\bar{x}\sqrt{N} \tag{5.46}$$

From (5.45) and (5.46), we get the following expression:

$$\bar{x} = \frac{1.5\kappa}{\sqrt{N}} \tag{5.47}$$

The obtained expression implies that the mean free path of a tribon is determined by not only the 'mass density' of the matter, but also the purely quantum parameters presented in (5.47) as the wave numbers κ. For numerical estimation of the coefficient of internal friction η, we will use the Newton's law for internal friction [69]:

$$\eta = \frac{1}{3}nm\theta\bar{x} \tag{5.48}$$

where n is the medium density; m is the particles mass (or quasiparticles effective mass); θ is the particles velocity; $\bar{x}$ is the mean free path.

We regard the product $m\theta$ as the momentum that a particle or quasiparticle has. In accordance with de Broglie equation, it can be 'assigned' the wave properties: $m\theta = \hbar\kappa$. If we assume $n = N/V$ and $V = \bar{x}^3$, (5.48) takes the following 'quasi-quantum' form:

$$\eta = \frac{1}{3}\frac{\hbar\kappa N}{\bar{x}^2} \tag{5.49}$$

If we denote the quantity $\hbar\kappa N$ as p_m, adding the average momentum to it, which is transmitted to macroobjects of the frictional contact (macroimpulse) substances and $\bar{x}^2$ is an area that is logical to identify with the dispersion cross-section σ, (5.49) may be rewritten in the following way:

$$\eta = \frac{1}{3}\frac{p_m}{\sigma} \tag{5.50}$$

(5.50) keeps its 'impulse' nature of the friction coefficient that appeared in (5.35), but now for macrosystems and macromotions. Friction, according to (5.50), is also caused by the impact of 'negative impulses' p_m from macrovolumes of the substance of the frictional contact medium. The source of these negative impulses is a flow of a big number of tribons of effective mass m, N in number. The action of this flow weakens with strengthening of tribons dispersion – the process characterized by the parameter σ, which is seen from (5.50).

Considering (5.35), (5.46) and (5.50), we may establish the correlation between various parameters that serve as numeric characteristics of a quantum tribosystem:

$$\sigma\omega\sqrt{N} = 0.5 \cdot 10^8 \text{ m/s}$$

or

$$\eta = \frac{1}{3}\frac{\hbar\omega}{c}n\sqrt{\frac{\sigma}{\pi}} = 10^{-5}\sqrt{\sigma\omega} \tag{5.51}$$

Comparing the latter expression with (5.36), we may get the following:

$$\gamma = 10^{-5}\sqrt{\frac{\sigma}{N}}\omega \tag{5.52}$$

The coefficient of friction γ is linked with the coefficient f from (5.26) and therefore, with the wave function that describes the state of the quantum tribosystem. The monograph [37] shows the correlation between the quantity ψ and the coefficient of friction. This correlation is determined by the passivation processes of the friction surfaces and the related appearance of lubrication films. This process results in the appearance of heterogeneous layer structures that provide the implementation of one of the most important principles in tribology—the principle of positive gradient of mechanical properties.

Definition 5.2. *Gradient* of mechanical properties is a vector that characterizes the changes in the mechanical properties of a surface element normal to it. Positive gradient of mechanical properties provides the concentration of shear deformations in the thin surface layer at external friction, thus keeping the continuity of the great bulk of friction material. The small value of shear deformation resistance that appears with that defines the minimum of the friction force and the magnitude of wear in a friction unit [3].

The quantum mechanical equation (5.52) that describes the behavior of the macrocoefficient of friction γ reflects the fact that the frictional interaction at micro- and nanolevels is considered using the formalism of quantum mechanics in the Schrödinger picture (Section

5.2). The comparison of the two pictures used in physics, the Heisenberg and Schrödinger pictures, described in Digression 5.1, allows to distinguish a formalized approach to the creation of calculation formalism of the theory. We see here two different views on fundamental mechanisms of the evolution of physical systems. In the Heisenberg picture, the kinetics of a quantum system are determined by the changes in space matrix. In the Schrödinger picture, motion was characterized by the changes in the physical system itself, caused by temporal dependency of the wave function ψ.

Representing the coefficient of friction γ as integral of motion and assigning all temporal dependencies to the tribons generating 'area', correlating quantum frictional equation with the Schrödinger picture, we do not exclude a dependence of the basis of Hilbert space where the tribosystem is on time. Therefore, the acts of tribotransformations should be inspired by spatial transformations.

In the Heisenberg picture, the bases of Hilbert spaces of tribosystems conform to external changes, e.g. they are linked with the action of a quantum tribometer (Def. 3.29) and the initial assumption that correlates the decoherence of properly tribosystem in non-local quantum state with its 'materialization' as a manifested tribosystem corresponds to the transition from infinite-dimensional Hilbert space to four-dimensional Minkowski space.

5.4 Evaluation of the Degree of Quantum Entanglement between Various Tribosystems

The quantitative estimation of the degree of entanglement, which arises in a tribosystem as it evolves, was done in [77]. This estimation was based on an assumption that each act of quantum measurement on the initial properly tribosystem leads to its decoherence, resulting in 'materialization' of the latter in 'our' world, four-dimensional Minkowski space [45]. Then, from the formal point of view, each measurement of properly tribosystem should 'generate' the state 4-vector $\left|\psi\right\rangle$; if there are n such measurement, they form a system of equations in the following form:

$$
\begin{aligned}
\left|\psi_1\right\rangle &= a_{11}\left|q_1\right\rangle + a_{12}\left|q_2\right\rangle + a_{13}\left|q3\right\rangle + a_{14}\left|q_4\right\rangle \\
\left|\psi_2\right\rangle &= a_{21}\left|q_1\right\rangle + a_{22}\left|q_2\right\rangle + a_{23}\left|q3\right\rangle + a_{24}\left|q_4\right\rangle \\
&\cdots\cdots\cdots\cdots\cdots\cdots\cdots\cdots\cdots\cdots \\
\left|\psi_n\right\rangle &= a_{n1}\left|q_1\right\rangle + a_{n2}\left|q_2\right\rangle + a_{n3}\left|q3\right\rangle + a_{n4}\left|q_4\right\rangle
\end{aligned}
\tag{5.53}
$$

where a_{ij} is the coefficient of vector expansion of the state vector; $\left|\psi_i\right\rangle$, $\left|q_i\right\rangle$ are the wave vectors that describe the tribosystem after the decoherence and the basic wave vectors, respectively.

The requirement for the wave functions that describe the quantum system state to be orthonormal leads to vanishing of the off-diagonal coefficient a_{ij} in the density matrix, if their indices differ by more than one. Then the density matrix of a tribosystem takes the form (1.3). [77, 78] assumed that the structure of the matrix (1.3) reflects the properties of Hilbert space and the features of the structure (if this term is applicable to this state of matter) properly tribosystem.

The (1.3) matrix form implies that a tribosystem in its non-local quantum state is a sixteen-dimensional object that includes three four-dimensional subsystems, which, under the hypotheses stated in [78], are potentially manifested tribosystems. By the analogy proposed by S. Doronin [38], they may be regarded as still invisible pictures from an undeveloped photo film. Therefore, between properly tribosystem and the manifested tribosystem, like between diverse variants of the decoherence of properly tribosystem, quantum correlations should exist. The diagonal elements of the density matrix describe the basic state of a quantum system and the off-diagonal elements characterize the correlations inside a tribosystem. Therefore, to estimate the contribution of quantum correlations to tribosystems development, it is necessary to get the explicit form of the density matrix.

A tribosystem, with its decoherence determined by the tribometer exposure on the original information field, 'manifesting itself' in our four-dimensional Minkowski world like all materialized physical systems, has a whole set of real-energy parameters, including positive entropy production. But the initial non-local quantum system (the reality), "...from which our dense world is projected... the world that has no mass and energy flows at all. It's a void, which nonetheless contains the whole set of classical (creative) energies in non-local superpositions.

All creative energies seem to compensate each other and, aggregated, form the all-embracing void..." [38]. That all-embracing void has no entropy and properly time, the direction of which is defined by entropy. The birth of a system with positive energy as a result of the decoherence, as stated in Section 1.4, predefines the appearance of its 'mirror reflection' at the same time, with negative entropy and different flow of time. To such systems, the approach is borrowed when considering a traditional EPR pair: two fermions with opposite directed spins [77]. Then let the state vector, which we denote as the ket-vector $\left|0\right\rangle$, corresponds to the system with positive entropy, $\left|1\right\rangle$ to the system with negative entropy. This EPR doublet should be described by state vectors $\left|01\right\rangle = \left|0\right\rangle\left|1\right\rangle$ and $\left|10\right\rangle = \left|1\right\rangle\left|0\right\rangle$, respectively. The state vector of this quantum doublet ψ equals

$$\left|\psi_\pm\right\rangle = \frac{1}{\sqrt{2}}\left(\left|01\right\rangle \pm \left|10\right\rangle\right) \tag{5.54}$$

In matrix form, these ket-vectors may be written this way: $\left|0\right\rangle = \begin{Vmatrix}1\\0\end{Vmatrix}$, $\left|1\right\rangle = \begin{Vmatrix}0\\1\end{Vmatrix}$, then:

$$\left|01\right\rangle = \left|0\right\rangle\left|1\right\rangle = \begin{Vmatrix}0\\1\\0\\0\end{Vmatrix}, \left|10\right\rangle = \left|1\right\rangle\left|0\right\rangle = \begin{Vmatrix}0\\0\\1\\0\end{Vmatrix} \tag{5.55}$$

We substitute the result of (5.55) in (5.54) and write the matrix form of the vectors $\left|\psi_+\right\rangle$ and $\left|\psi_-\right\rangle$:

$$\begin{gathered}\left|\psi_+\right\rangle = \frac{1}{\sqrt{2}}\left(\left|01\right\rangle + \left|10\right\rangle\right) = \frac{1}{\sqrt{2}}\left(\begin{Vmatrix}0\\1\\0\\0\end{Vmatrix} \pm \begin{Vmatrix}0\\0\\1\\0\end{Vmatrix}\right) = \frac{1}{\sqrt{2}}\begin{Vmatrix}0\\1\\1\\0\end{Vmatrix}\\ \left|\psi_-\right\rangle = \frac{1}{\sqrt{2}}\left(\left|01\right\rangle - \left|10\right\rangle\right) = \frac{1}{\sqrt{2}}\left(\begin{Vmatrix}0\\1\\0\\0\end{Vmatrix} - \begin{Vmatrix}0\\0\\1\\0\end{Vmatrix}\right) = \frac{1}{\sqrt{2}}\begin{Vmatrix}0\\1\\-1\\0\end{Vmatrix}\end{gathered} \tag{5.56}$$

The state vector of the manifested tribosystem is a superposition of the ket-vectors $\left|psi_+\right\rangle$ and $\left|\psi_-\right\rangle$:

$$\Big|\psi\Big\rangle = a\Big|\psi_+\Big\rangle + b\Big|\psi_-\Big\rangle \tag{5.57}$$

Accordingly, the bra-vector $\Big\langle\psi\Big|$ will be written as

$$\Big\langle\psi\Big| = a^*\Big\langle\psi_+\Big| + b^*\Big\langle\psi_-\Big| \tag{5.58}$$

The coefficients of vector expansion in (5.57), (5.58) have the following properties: $a^2 + b^2 = 1$, $a^2 = aa^*$, $b^2 = bb^*$, where a^* and b^* are complex conjugate quantities a and b. Let us write the projector $\Big|\psi\Big\rangle\Big\langle\psi\Big|$:

$$\Big|\psi\Big\rangle\Big\langle\psi\Big| = a^2\Big|\psi_+\Big\rangle\Big\langle\psi_+\Big| + ab^*\Big|\psi_+\Big\rangle\Big\langle\psi_-\Big| + ba^*\Big|\psi_-\Big\rangle\Big\langle\psi_+\Big| + b^2\Big|\psi_-\Big\rangle\Big\langle\psi_-\Big| \tag{5.59}$$

The ‘small projectors’ included in (5.59) will be equal to:

$$\begin{aligned}
\Big|\psi_+\Big\rangle\Big\langle\psi_+\Big| &= \frac{1}{2}\Big(\Big|01\Big\rangle\Big\langle 01\Big| + \Big|10\Big\rangle\Big\langle 10\Big| + \Big|01\Big\rangle\Big\langle 10\Big| + \Big|10\Big\rangle\Big\langle 01\Big|\Big)\\
\Big|\psi_+\Big\rangle\Big\langle\psi_-\Big| &= \frac{1}{2}\Big(\Big|01\Big\rangle\Big\langle 01\Big| - \Big|01\Big\rangle\Big\langle 10\Big| + \Big|10\Big\rangle\Big\langle 01\Big| - \Big|10\Big\rangle\Big\langle 10\Big|\Big)\\
\Big|\psi_-\Big\rangle\Big\langle\psi_+\Big| &= \frac{1}{2}\Big(\Big|01\Big\rangle\Big\langle 01\Big| + \Big|01\Big\rangle\Big\langle 10\Big| - \Big|10\Big\rangle\Big\langle 01\Big| - \Big|10\Big\rangle\Big\langle 10\Big|\Big)\\
\Big|\psi_-\Big\rangle\Big\langle\psi_-\Big| &= \frac{1}{2}\Big(\Big|01\Big\rangle\Big\langle 01\Big| - \Big|01\Big\rangle\Big\langle 10\Big| - \Big|10\Big\rangle\Big\langle 01\Big| + \Big|10\Big\rangle\Big\langle 10\Big|\Big)
\end{aligned} \tag{5.60}$$

Calculate the ‘basic’ projectors:

$$\Big|01\Big\rangle\Big\langle 01\Big| - \begin{Vmatrix}0\\1\\0\\0\end{Vmatrix} \times \begin{Vmatrix}0 & 1 & 0 & 0\end{Vmatrix} = \begin{Vmatrix}0&0&0&0\\0&1&0&0\\0&0&0&0\\0&0&0&0\end{Vmatrix}$$

$$\Big|10\Big\rangle\Big\langle 10\Big| = \begin{Vmatrix}0&0&0&0\\0&0&0&0\\0&0&1&0\\0&0&0&0\end{Vmatrix}, \Big|01\Big\rangle\Big\langle 10\Big| = \begin{Vmatrix}0&0&0&0\\0&0&1&0\\0&0&0&0\\0&0&0&0\end{Vmatrix}, \tag{5.61}$$

$$\Big|01\Big\rangle\Big\langle 01\Big| = \begin{Vmatrix}0&0&0&0\\0&0&0&0\\0&1&0&0\\0&0&0&0\end{Vmatrix}$$

By substituting the value of the 'basic' projectors (5.61) in (5.60), we get:

$$\left|\psi_+\right\rangle\left\langle\psi_+\right| = \frac{1}{2}\left(\begin{Vmatrix}0&0&0&0\\0&1&0&0\\0&0&0&0\\0&0&0&0\end{Vmatrix}+\begin{Vmatrix}0&0&0&0\\0&0&0&0\\0&0&1&0\\0&0&0&0\end{Vmatrix}+\begin{Vmatrix}0&0&0&0\\0&0&1&0\\0&0&0&0\\0&0&0&0\end{Vmatrix}+\begin{Vmatrix}0&0&0&0\\0&0&0&0\\0&1&0&0\\0&0&0&0\end{Vmatrix}\right)$$

$$= \frac{1}{2}\begin{Vmatrix}0&0&0&0\\0&1&1&0\\0&1&1&0\\0&0&0&0\end{Vmatrix} = \begin{Vmatrix}0&0&0&0\\0&1/2&1/2&0\\0&1/2&1/2&0\\0&0&0&0\end{Vmatrix}$$

Hence,

$$\left|\psi_+\right\rangle\left\langle\psi_-\right| = \begin{Vmatrix}0&0&0&0\\0&1/2&-1/2&0\\0&1/2&-1/2&0\\0&0&0&0\end{Vmatrix}, \left|\psi_-\right\rangle\left\langle\psi_+\right| = \begin{Vmatrix}0&0&0&0\\0&1/2&1/2&0\\0&-1/2&-1/2&0\\0&0&0&0\end{Vmatrix}$$

$$\left|\psi_-\right\rangle\left\langle\psi_-\right| = \begin{Vmatrix}0&0&0&0\\0&1/2&-1/2&0\\0&-1/2&1/2&0\\0&0&0&0\end{Vmatrix} \tag{5.62}$$

By substituting the projectors (5.62) in (5.59), we get:

$$\left|\psi\right\rangle\left\langle\psi\right| = \begin{Vmatrix}0&0&0&0\\0&a_1&a_2&0\\0&a_3&a_4&0\\0&0&0&0\end{Vmatrix} \tag{5.63}$$

where

$$a_1 = \frac{1}{2}\left[a\left(a-b^*\right)-b\left(b-a^*\right)\right], a_2 = \frac{1}{2}\left[a\left(a+b^*\right)+b\left(b+a^*\right)\right]$$

$$a_3 = \frac{1}{2}\left[a\left(a-b^*\right)+b\left(b-a^*\right)\right], a_4 = \frac{1}{2}\left[\left(a+b^*\right)-b\left(b+a^*\right)\right]$$

Solving the equation of the form [77], we find the eigensolutions for the coefficients of the density matrix λ:

$$\begin{vmatrix} -\lambda & 0 & 0 & 0 \\ 0 & a_1 - \lambda & a_2 & 0 \\ 0 & a_3 & a_4 - \lambda & 0 \\ 0 & 0 & 0 & 0 \end{vmatrix} = 0 \tag{5.64}$$

which reduces to the quadratic equation: $\lambda^2 - (a_1 + a_4)\lambda - (a_1 a_2 - a_2 a_3) = 0$. The roots of this equation are following:

$$\begin{aligned} \lambda_1 &= \frac{a^2 - b^2}{2} + \sqrt{\frac{(a^2 - b^2)^2 + 8a^3 b}{4}} \\ \lambda_2 &= \frac{a^2 - b^2}{2} - \sqrt{\frac{(a^2 - b^2)^2 + 8a^3 b}{4}} \end{aligned} \tag{5.65}$$

Since the coefficients a and b, under their physical meaning, should be greater than one, we may neglect their higher degrees. Then (5.65) is simplified to the form:

$$\begin{aligned} \lambda_1 &= 0 \\ \lambda_2 &= a^2 - b^2 = 2a^2 - 1 \end{aligned} \tag{5.66}$$

The parameters λ_1 and λ_2 are the eigenvalues of the density matrix, by which the degree of entanglement between the components of properly tribosystem can be evaluated. As first approximation, it had been done in [77], but some of the conclusions from there need to be specified and the definition extended. With the eigenvalue $\lambda_1 = 0$, the value of the matching parameter from (1.1) also equals zero. Substituting these values in (1.2) to define the measure of entanglement, we get a meaningless expression that contains a logarithm of zero. Therefore, the solution that corresponds to zero eigenvalues of the density matrix may be discarded. If we consider the solution λ_1 in (5.66), based on (1.1) and the relations between the coefficients of vector expansion of the state vector, we may show that $c_{1,2}^2 = 1 \pm 4b^2$. Then the entanglement $E(\psi)$ will be represented by the following relation:

$$E(\psi) = -\left(1 - 2b^2\right) \log_2 \left[\frac{1}{2}(1 - 2b^2)\right] \tag{5.67}$$

Hence, the entanglement value depends on the quantity b^2; in fact, the quantum system probability described by the ket-vector $\left|\psi_-\right\rangle$. The quantity b^2 changes in range 0..1. Let us calculate the entanglement value for the extreme values of the parameter b^2. If $b = 0$, then $E(\psi)$ equals 1, that is, a system in one single state is entangled with itself. In this sense, the result may be considered trivial. Provided the value of the logarithmic function is taken in (5.67) by absolute value, the second 'pure' system at $b = 1$ is also absolutely entangled with itself: $E(\psi) = 1$. Nonetheless, passing points, it is advisable to introduce a requirement that additionally limits the values of the coefficients of vector expansion and, at the same time, does not allow negative values of the logarithms arguments, i.e. $(1 - b^2) > 0$, or

$$b^2 < 0.5 \tag{5.68}$$

From the condition (5.68), to estimate the intermediate values of entanglement, we arbitrarily choose values for the coefficients b, e.g. $b_1^2 = 0.4, b_2^2 = 0.1$. Then, disregarding the meaningless variants of the entanglement parameter values, we get:

$$\begin{aligned} E(\psi)|_{b_1} &= 0.469 \\ E(\psi)|_{b_2} &= 0.087 \end{aligned} \tag{5.69}$$

We may represent the quantum correlation parameter between manifested and properly tribosystems (Fig. 5.2).

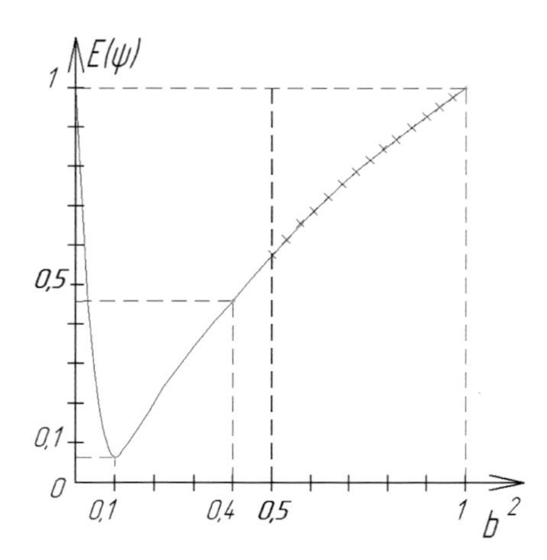

Figure 5.2: The dependence of tribosytems entanglement and the features of their quantum structure (the shadowed area of the curve corresponds to the violation of the condition (5.68))

The dependence $E(\psi)$ on the coefficients of vector expansion a and b allows to establish the impact of the tribosystem quantum structure on the value quantum exchange interactions. Thereupon it is important to study the possibilities of quantum correlations between various 'generations' of the decoherences – the manifested tribosystems whose existence is seen from the structural features of the density matrix (1.3). In accordance with the assumption stated in [77], from the structure of the matrix (1.3) follow three possible variants of 'materialization' of the tribosystem because of the decoherence:

$$A = \begin{Vmatrix} a_{11} & a_{12} \\ a_{21} & a_{22} \end{Vmatrix}, B = \begin{Vmatrix} a_{22} & a_{23} \\ a_{32} & a_{33} \end{Vmatrix}, C = \begin{Vmatrix} a_{33} & a_{34} \\ a_{43} & a_{44} \end{Vmatrix} \tag{5.70}$$

The subsystems presented in matrices (5.70) exist in the objective reality and are very probably connected by the 'forces' of quantum correlations (which, in particular, are apparent from the very structure of the matrix (1.3)). They are separable quantum objects and the wave function of the manifested tribosystem, according to [38], may be represented as a tensor product of partial-wave functions:

$$\psi_T = \psi_A \otimes \psi_B \otimes \psi_C \tag{5.71}$$

where ψ_A, ψ_B, ψ_C are partial-wave function of the tribosystem (5.70).

The symmetric form of the density matrix (1.3) reduces the consideration of quantum entanglement between the subsystems A, B and C to the study of the interactions of the doublets $A - B, B - C$, which should have the same form. Therefore, an analysis inside one subsystem pair is absolutely equivalent for another doublet. Each subsystem A, B, C appears at a definite moment of quantum measurement of properly tribosystem, in accordance with the rule: one measurement – reduction – decoherence – one manifested tribosystem [78]. These tribosystems (subsystems) 'materialize' in Minkowski continuum, the spatial coordinates of which, owing to homogeneity and isotropy, are completely equal quantities [77, 78]. Therefore, the 'localization' of the manifested tribosystem for simplicity may be fixed with one coordinate r and the entire continuum may be regarded as a 'plane' described by two coordinates: matrix r and temporal t. Then the act of birth of one of the subsystems A, B, C at a definite moment may be written in the state vector form:

$$\left|+\right\rangle = \begin{Vmatrix} 1 \\ x \end{Vmatrix} \tag{5.72}$$

If for some reason no birth occurred, this vector takes the form:

$$\Big|-\Big\rangle = \begin{Vmatrix} 0 \\ x \end{Vmatrix} \tag{5.73}$$

The state vectors of subsystem doublets $A - B$ and $B - C$ are obtained by transformations like (1.9)–(1.14) [77]:

$$\Big|+-\Big\rangle = \begin{Vmatrix} 0 \\ x \\ 0 \\ x^2 \end{Vmatrix}, \Big|-+\Big\rangle = \begin{Vmatrix} 0 \\ 0 \\ x \\ x^2 \end{Vmatrix}, \Big|\psi_+\Big\rangle = \frac{1}{\sqrt{2}}\begin{Vmatrix} 0 \\ 0 \\ x \\ 2x^2 \end{Vmatrix}, \Big|\psi_-\Big\rangle = \frac{1}{\sqrt{2}}\begin{Vmatrix} 0 \\ x \\ -x \\ 0 \end{Vmatrix}$$

$$\Big|\psi\Big\rangle\Big\langle\psi\Big| = a^2\Big|\psi_+\Big\rangle\Big\langle\psi_+\Big| + b^2\Big|\psi_-\Big\rangle\Big\langle\psi_-\Big| + ab^*\Big|\psi_+\Big\rangle\Big\langle\psi_-\Big| + ba^*\Big|\psi_-\Big\rangle\Big\langle\psi_+\Big| \tag{5.74}$$

Let us write the 'internal' projectors in the last equation:

$$\Big|\psi_+\Big\rangle\Big\langle\psi_+\Big| = \frac{1}{2}\Big\{\Big|+-\Big\rangle\Big\langle+-\Big| + \Big|-+\Big\rangle\Big\langle-+\Big| + \Big|+-\Big\rangle\Big\langle-+\Big| + \Big|-+\Big\rangle\Big\langle+-\Big|\Big\} \tag{5.75}$$

The expression (5.75) can be simplified if we use the 'selection rule', which S.I. Doronin formulated as follows: "Speaking in simple words, to get a reduced density matrix, we need to leave only those matrix elements at which there are equal figures at a given position... The other matrix elements vanish" [40, 77]. Applying this rule, the projector (5.75) is simplified to the form:

$$\Big|\psi_+\Big\rangle\Big\langle\psi_+\Big| = \frac{1}{2}\Big\{\Big|+-\Big\rangle\Big\langle+-\Big| + \Big|-+\Big\rangle\Big\langle-+\Big|\Big\} \tag{5.76}$$

Similarly, we get the expression for other projections in (5.74):

$$\begin{aligned} \Big|\psi_-\Big\rangle\Big\langle\psi_-\Big| &= \frac{1}{2}\Big\{\Big|+-\Big\rangle\Big\langle+-\Big| + \Big|-+\Big\rangle\Big\langle-+\Big|\Big\} = \Big|\psi_+\Big\rangle\Big\langle\psi_+\Big| \\ \Big|\psi_+\Big\rangle\Big\langle\psi_-\Big| &= \frac{1}{2}\Big\{\Big|+-\Big\rangle\Big\langle+-\Big| - \Big|-+\Big\rangle\Big\langle-+\Big|\Big\} = \Big|\psi_-\Big\rangle\Big\langle\psi_+\Big| \end{aligned} \tag{5.77}$$

The expressions may be rewritten in matrix form:

$$\Big|\psi_+\Big\rangle\Big\langle\psi_+\Big| = \Big|\psi_-\Big\rangle\Big\langle\psi_-\Big| = \frac{1}{2}\begin{Vmatrix} 0 & 0 & 0 & 0 \\ 0 & x^2 & 0 & x^3 \\ 0 & 0 & x^2 & x^3 \\ 0 & x^2 & x^3 & 2x^4 \end{Vmatrix}$$
$$\Big|\psi_+\Big\rangle\Big\langle\psi_-\Big| = \Big|\psi_-\Big\rangle\Big\langle\psi_+\Big| = \frac{1}{2}\begin{Vmatrix} 0 & 0 & 0 & 0 \\ 0 & x^2 & 0 & x^3 \\ 0 & 0 & -x^2 & -x^3 \\ 0 & x^3 & -x^3 & 0 \end{Vmatrix} \quad (5.78)$$

Substituting the value of (5.78) in (5.74), we get the following matrix formula of the projector $\Big|\psi\Big\rangle\Big\langle\psi\Big|$:

$$\Big|\psi\Big\rangle\Big\langle\psi\Big| = \begin{Vmatrix} 0 & 0 & 0 & 0 \\ 0 & Y_1 & 0 & Y_1 \\ 0 & Y_1 & Y_2 & Y_2 \\ 0 & Y_2 & Y_2 & 1 \end{Vmatrix} \quad (5.79)$$

where $Y_1 = \dfrac{1}{2}\Big\{1 + \Big(ab^* + b * a\Big)\Big\}$, $Y_2 = \dfrac{1}{2}\Big\{1 - \Big(ab^* + b * a\Big)\Big\}$, provided the moment x is one.

Consider the secular equation of the form (5.64):

$$\begin{vmatrix} -\lambda & 0 & 0 & 0 \\ 0 & Y_1 - \lambda & 0 & Y_1 \\ 0 & Y_1 & Y_2 - \lambda & Y_1 \\ 0 & Y_1 & Y_2 & 1 - \lambda \end{vmatrix} = 0$$

Or:

$$\lambda\left\{(Y_1 - \lambda)\begin{vmatrix} Y_2 - \lambda & Y_2 \\ Y_2 & 1 - \lambda \end{vmatrix} + Y_1 \begin{vmatrix} Y_1 & Y_2 - \lambda \\ Y_1 & Y_2 \end{vmatrix}\right\} = 0 \quad (5.80)$$

The trivial solution λ_0 of (5.80) is a zero value of the parameter λ. To look for other solutions, consider the zero value of the expression in curly brackets, which is transformed to the form [77]:

$$-\lambda^2 + \lambda(1 + Y_1 + Y_2) - \lambda(Y_1^2 + Y_1 + Y_1 Y_2 + Y_2) + Y_1 Y_2 = 0 \quad (5.81)$$

Substituting in (5.81) the values of the parameters Y_1, Y_2 of the determinant of (5.79) and neglecting higher degrees at the coefficients

a, b, a^*, b^*, we obtain a comfortable analysis form of cubic equation with constant coefficients:

$$\lambda^3 - 2\lambda^2 + 1.5\lambda - 0.5 = 0 \tag{5.82}$$

This equation is similar to cubic equations:

$$ax^3 + bx^2 + cx + d = 0 \tag{5.83}$$

The parameters a, b, c, d and the unknown x from (5.83) do not coincide with the quantum parameters that we denoted by the same letters. (5.83) is solved by substitution [51, 77]:

$$x = y - \frac{b}{3a} \tag{5.84}$$

It can be seen that a new unknown y appears; with it the equation (5.83) is transformed to a simpler form:

$$y^3 + py + q = 0 \tag{5.85}$$

where $p = \frac{c}{a} - \frac{1}{3}\left(\frac{b}{a}\right)^2$, $q = \frac{2}{27}\left(\frac{b}{a}\right)^2 - \frac{1}{3}\frac{bc}{a^2} + \frac{d}{a}$

If the discriminant of the equation (5.85) $\Delta = \frac{p^3}{27} + \frac{q^2}{4}$ is greater than zero, the equation has one real and two complex roots. If the discriminant is less than zero, all the three roots of (5.85) will be real. Let us substitute the values of a, b, c, d from (5.82) into (5.85) and estimate its discriminant. The calculation shows that $\Delta = 0.36$, i.e. the discriminant is greater than zero and (5.82) has one real and two complex roots and owing to their physical meaning, we do not need to consider. Under [51, 77], the only solution of this equation has the form:

$$y = u + \theta \tag{5.86}$$

where $u = \sqrt{-0.5q^2 + \sqrt{\Delta}}$, $\theta = \sqrt{-0.5q^2 - \sqrt{\Delta}}$

Substituting the numeric relations from (5.82)–(5.85), we get $u = 1.1, \theta = 0$. Therefore, the only root of (5.85) is equal to 1.1. From (5.84) we may calculate the value of λ; it equals 0.43, which gives the consistency value $c = 0.82$ and the measure of entanglement $E(\psi) = 0.745$. The values found show a considerable quantum correlation between manifested tribosystems A, B, C. The original equation (5.80) has one more solution that was called trivial and the existence of which is not considered in purely mathematical analysis. However, the physical reality of quantum interactions between systems indicate the need to consider those variants as well. In case of $\lambda = 0$, the consistency parameter calculated from (1.1) also equals zero, which gives a meaningless result from the viewpoint of mathematics $E(\psi)$. Therefore, this solution variant was not considered. The theoretically discovered quantum correlations between manifested tribosystems match the above Everett concept that explains the nature of the wave function reduction. Upon this view, one may assume that the manifested tribosystems in the form of subsystems of original properly tribosystem exist 'simultaneously', being in different subspaces of sixteen-dimension phase space of initial properly tribosystem (1.3). The above calculation indicates a high probability of the existence of quantum entanglement non-local relation between the subsystems of the tribosystem. This relation between the manifested tribosystems A, B, C is almost twice more 'strong' as compared to their connections to the initial properly tribosystem (Fig. 5.3).

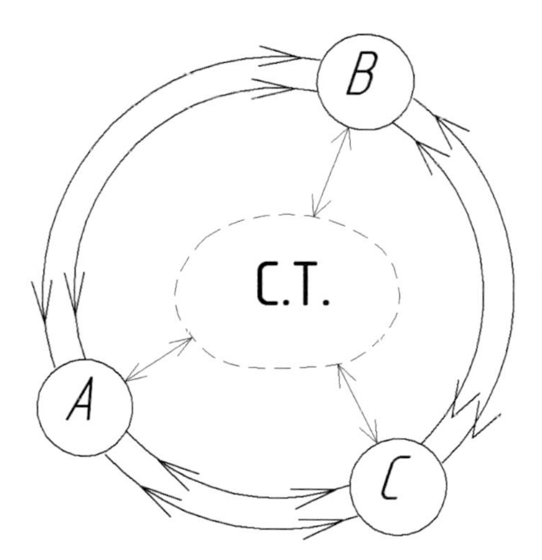

Figure 5.3: Sketch of quantum entanglement between properly tribosystems (P.T.) and manifested tribosystems A, B, C.

This graphical representation of quantum objects is more than conventional and the attempts at such representations are questionable. That is why the amount of illustrations in [78] is very modest. Nonetheless, the desire to introduce graphical images into quantum description of physical systems arises from time to time. The examples are the famous Feynman diagrams [54], as well as the attempts to draw non-local quantum relations with Bloch sphere [38].

Digression 5.2. Bloch sphere is an attempt to graphically represent the entangled quantum objects in non-local information state. If a physical system is in its pure state, i.e. is closed, the measure of information in it is maximal and equal to one. If a physical system is open, i.e. is in mixed state, the information in it are fractions of one equal to 2^{-N}, where $2N$ is the dimension of the Hilbert space where this physical system is located. The unit of measurement of quantum information is q-bit. As noted above, any state vector may be considered consisting of elementary vector, q-bits [38]. The density matrix of a single q-bit may be represented by a point in three-dimensional space; that is, there is a bijection between the density matrix and the point on the sphere of unitary radius. For the pure state, these points belong to the surface of the sphere. Under interaction with the environment, i.e. decoherence, the state vector 'subsides' under the surface of the sphere. Such a physical system is called the Bloch sphere. Hence, the deep points of that sphere describe related systems and their determining vectors do not describe ideal circles but something close to ellipsoids. In an extreme case, when the state of q-bit becomes mixed, the Bloch sphere space contracts up to a segment on the quantization axis (z-axis) between the values $1/2, -1/2$; that is, the minimum that remained from the q-bit (Figs. 5.4 and 5.5).

The Bloch sphere visually represents how our classical world correlates with the quantum reality ('becomes embedded in it') [38] and how this classical reality arises as a result of decoherence. Let our physical system take only two states – above and below along the quantization axis z. The points on that axis within the Bloch sphere are classical states of a physical system manifested at the decoherence and forms the so-called 'classical domains'. The classical domain takes a small part of all the possible states of the quantum system (Fig. 5.5). The rest of the Bloch sphere volume is the quantum domain. According to S. Doronin: "The entire world with all its matter, substance and physical fields is only a single smallest and completely insignificant point on the z-axis. Many consider everything the base of the universe and the only objective reality that exists is only a projection in quantum reality, a 'pale shadow' that falls from the state vector onto the quantization axis with the evolution of a more complex joint quantum reality..." [38].

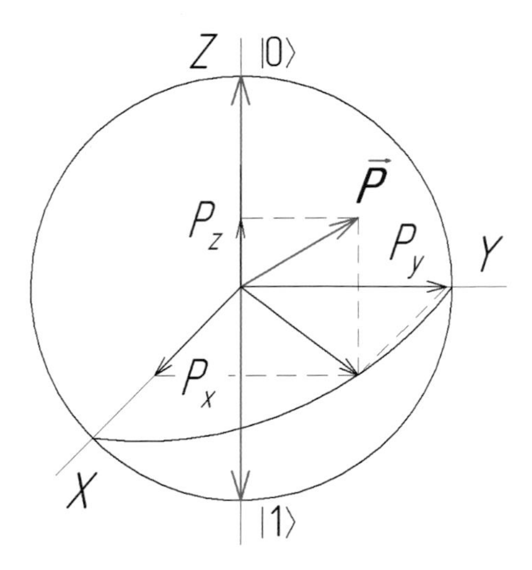

Figure 5.4: Sketch of the Bloch sphere: $\vec{P}$ is the Bloch vector [38]

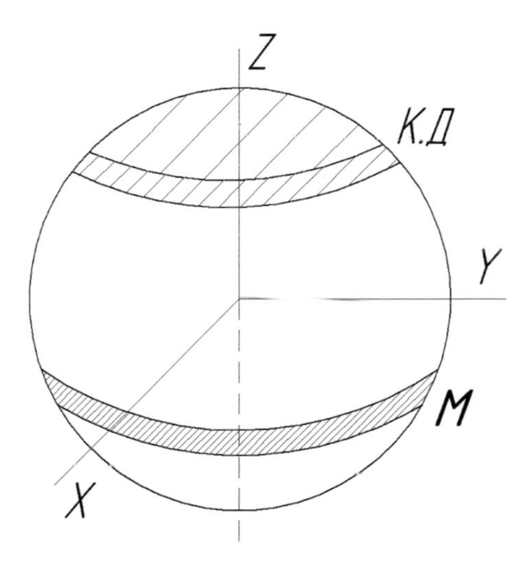

Figure 5.5: Formation of the classical domain (C.D.). M is the maximum entanglement area

The points on the Bloch sphere surface corresponding to the pure state are localized near the poles of the system; the other points correspond to mixed quantum states (Fig. 5.5). Of non-local pure states, one may pick out those that correspond to the 'equator' point on the Bloch sphere that serves, to some extent, the antipodes of polar states of the sphere. If on the poles all the systems have definite macrostates, gradually 'diffusing' with the motion inside the Bloch sphere, the system on the equator is in an equally probable state of 'alive-dead' Schrödinger cat that corresponds to the stage of maximum entanglement of all its parts.

Epilogue

Section 5.4 in its contents resembles the educational articles in modern physics. From the authors' point of view, it is unknown whether it is bad or good. We aimed to show the general picture of changes in a tribosystem structure, using full sets of visual representations found in scientific and educational literature by L. Cooper, R. Feynman, K. Thorne, S. Hawking et al. We hope that the final chapter fully completes the general picture of fundamental quantum-relativistic mechanism of tribosystem evolution while preserving the elements of rigor. Thinking about the graphical forms which could represent the nuances of the quantum world to a small extent, we suddenly discovered the 'implicit graphicity' of the matrix notation of quantum quantities. In presenting the material, the transposition of matrices (3.69) was presented a sequence of two rotations of coordinate axes by 45° (3.72), and Fig. 3.4, where the 'chess' structure of matrix (3.59) reflects not only the degenerate nature of the quantum system showing that its phase space contains three more subspaces representing Minkowski continuum, but also that the structure of matrix (3.59) contains the basic elements a_{22} and a_{33} that are common for two subspaces cells at once. The latter is very similar to the transitions between the phase subspaces as shown graphically in Fig. 5.4.

Anyway, the graphical representation of Section 5.4 is a summary of this book, in which we made an attempt to consistently solve one of the problems of theoretical tribology using the methods of modern physics – the only right way from our point of view.

The authors are thankful to everyone who opens this book, as well as to those who express their views on its contents.

Bibliography

[1] Jost H.P. "An analytical review of the tribology past and future". In: *Rus. J. of Friction and Wear* 11.1 (1990), pp. 149–159.

[2] *Lubrication (tribology) Education and Research (Jost Report). Department of Education and Science,* London: HMSO, 1966, 210 pp.

[3] *Slovar-spravochnik po treniyu, iznosu i smazke detaley mashin.* [Dictionary of Friction, Wear and Lubrication of Machine Parts]. Russian. Ed. by Zozula V.D. Shvedkov E.L. Rovinsky D.Y. Kiev: Naukova Dumka, 1979, 185 pp.

[4] *Spravochnik po tribotekhnike.* [Reference Book in Tribotechnics]. Russian. Ed. by Chichinadze A.V. Hebd M. Vol. 1. 3 vols. Moscow: Mashinostroyeniye, 1989, 400 pp.

[5] Myshkin N.K. Petrokovec M.I. *Tribologiya. Printsipy i prilozheniya.* [Tribology. Principles and Applications]. Russian. Gomel: IMMS, 2012, 310 pp.

[6] Bely V.A. Myshkin N.K. Ledem K., ed. *Tribologiya: Issledovaniya i prilozheniya: Opyt SShA i stran SNG.* [Tribology: Researches and Applications: USA and CIS Experience]. Russian. Moscow-New York: Mashinostroyeniye, Allerton Press, 1993, 454 pp.

[7] Sviridenok L.I. Myshkin N.K. et al. *Akusticheskiye i elektricheskiye metody v tribologii.* [Acoustic and Electric Methods in Tribology]. Russian. Minsk: Nauka i Tekhnika, 1987, 280 pp.

[8] Akhmatov A.S. *Molekularnaya fizika granichnogo treniya.* [Molecular Physics of Boundary Friction]. Russian. Moscow: Fiziko-matematicheskaya literatura, 1963, 471 pp.

[9] Yepifanov G.I. Minayev N.I. "Eksperimentalnoye sopostavleniye sil treniya i adgezii". [Experimental Comparison of Friction and Adhesion Forces]. Russian. In: *Izvestiya vuzov, ser. Fizika* 1 (1959), pp. 21–27.

[10] Cameron A. Ettles C.M. *Basic Lubrication Theory*. Chichister: Ellis Horwood Ltd., 1981, 304 pp.

[11] Deryagin B.V. *Chto takoye treniye?* [What is Friction?] Russian. Moscow: Academy of Sciences, USSR, 1963, 210 pp.

[12] Hardy W.B. *Collected Works*. Cambridge University Press, 1936, 375 pp.

[13] Bowden F.P. Tabor D. *Friction and Lubrication of Solids*. Clarendon Press, 2001, 374 pp.

[14] Dowson D. *History of Tribology*. 2nd ed. London: Professional Eng. Publ., 1998, 806 pp.

[15] Kragelsky I.V. *Treniye i iznos*. [Friction and Wear]. Russian. Moscow: Mashgiz, 1962, 384 pp.

[16] Kostecki B.I. *Poverkhnostnaya prochnost materialov pri trenii*. [Surface Strength of Materials in Friction]. Russian. Kiev: Tekhnika, 1976, 283 pp.

[17] Garkunov D.N. *Nauchnyye otkrytiya v tribotekhnike. Effekt bezyznosnosti. Vodorodnoye iznashivaniye*. [Scientific Discoveries in Tribotechnics. Wearless Effect. Hydrogen Wear]. Russian. Moscow: MSHA, 2004, 384 pp.

[18] Pitoyavsky L.P. "Makroskopicheskiye kvantovyye effekty". [Macroscopic Quantum Effects]. Russian. In: *Fizicheskaya Enciklopediya*. [Physics Encyclopaedia]. Ed. by Prokhorov A.M. Vol. 3. 5 vols. Moscow: Bolshaya Rossiyskaya Enciklopediya, 1992, pp. 29–31.

[19] Vasko F.T. Gribnikov Z.S. "Plazma tverdykh tel". [Plasma of Solids]. Russian. In: *Fizicheskaya Enciklopediya*. [Physics Encyclopedia]. Vol. 3. 5 vols. Moscow: Bolshaya Rossiyskaya Enciklopediya, 1992, pp. 600–604.

[20] Thiessen P.A. Meyer K. Heinicke G. "Grundlagen der Tribochemie". German. In: *Abn. Dtsch. Acad. Wiss. K1* 1 (1966), pp. 15–33.

[21] Nakayama K. "Triboemission and Triboplasma Generation with DLC Films". In: *Tribology of Diamond-like Carbon Films.* Berlin: Elsevier, 2008, pp. 291–310.

[22] Konyushaya Y.P. *Otkrytiya i nauchno-tekhnicheskaya revolutsiya.* [Discoveries and Scientific-Technical Revolution]. Moscow: Rabochiy, 1974, 496 pp.

[23] Schipp P.A. *Albert Einstein: Autobiographical Notes.* Open Court Pub Co., 1979, 95 pp.

[24] Feynman R. Leighton R. Sands M. *The Feynman Lectures on Physics.* Vol. 2, Addison-Wesley, 1964.

[25] Huggett N. *Zeno's Paradoxes.* Stanford Encyclopedia of Philosophy. Stanford University. 2010. URL: http://plato.stanford.edu/entries/paradox-zeno/.

[26] Barashenkov B.S. "Minus-materiya". [Minus-Matter]. Russian. In: *Znaniye-sila* 12 (1991), pp. 17–22.

[27] Hawking S. Mlodinow L. *The Grand Design.* New York: Bantam, 2010, 208 pp.

[28] Feygin O.O. *Velikaya kvantovaya revolutsiya.* [The Great Quantum Revolution]. Russian. Moscow: Eksmo, 2009, 256 pp.

[29] Feygin O.O. *Fizika nerealnogo.* [Physics of Unreal]. Russian. Moscow: Eksmo, 2010, 272 pp.

[30] Einstein A. Podolsky B. Rosen N. Bohr N. Fock V.A. *Can Quantum Mechanical Description of Physical Reality be Considered Complete?* Stanford University. 2013. URL: http://plato.stanford.edu/entries/qt-epr/.

[31] Landau L.D. Lifschitz E.M. *Kvantovaya mekhanika (nerelativistskaya teoriya).* [Quantum Mechanics (Non-Relativistic Theory)]. Vol. 3: *Teoreticheskaya fizika.* [Theoretical Physics]. Russian. Moscow: Nauka, 1974, 464 pp.

[32] Grib A.A. "Neravenstva Bella i eksperimentalnaya proverka kvantovykh korrelaciy na makroskopicheskikh rasstoyaniyakh". [Bell Inequalities and Experimental Check of Quantum Correlations at Microscopic Distances]. Russian. In: *Uspekhi fizicheskikh nauk. [Progress in Physical Sciences]* 142.4 (1984), pp. 619–634.

[33] Arsenov O.O. *Parallelnyye Vselennyye.* [Parallel Universes]. Russian. Moscow: Eksmo, 2011, 272 pp.

[34] Einstein A. Podolsky B. Rosen N. "Can quantum-mechanical description of physical reality be considered complete?" In: *Physical Review* 47 (1935), pp. 777–780.

[35] Bohr N. "Can quantum-mechanical description of physical reality be considered complete?" In: *Physical Review* 48 (1935), pp. 696–702.

[36] Spassky B.I. Moskovsky A.B. "O nelokalnosti v kvantovoy fizike". [On Non-locality in Quantum Physics]. Russian. In: *Uspekhi fizicheskikh nauk. [Progress in Quantum Physics]* 142.4 (1984), pp. 599–616.

[37] Feynman R. *Nobel Lecture*. Nobel Foundation. 1965. URL: `http://www.nobelprize.org/nobel_prizes/physics/laureates/1965/feynman-lecture.html`.

[38] Doronin S.I. *Kvantovaya magiya*. [Quantum Magic]. Russian. URL: `http://ligis.ru/librari/2698.htm`.

[39] Bargatin I.V. Grishanin B.A. Zadnov B.N. "Zaputannyye kvantovyye sostoyaniya atomnykh sistem". [Entangled Quantum States of Atomic Systems]. In: *Uspekhi fizicheskikh nauk. [Progress in Physical Sciences]* 171.6 (2001), pp. 625–646.

[40] Doronin S.I. "Mera kvantovoy zaputannosti chistykh sostoyaniy". [Measure of Pure States Quantum Entanglement]. Russian. In: *Kvantovaya magiya. [Quantum Magic]* 1.1 (2004), pp. 1123–1137.

[41] Walther P. Pan J.W. Aspelmeyer M. Zeilinger A. "De Broglie wavelength of a nonlocal four-photon state". In: *Nature* 429 (2004), pp. 158–161.

[42] Bouwmeester D. Ekert A. Zeilinger A. *The Physics of Quantum Information: Quantum Cryptography, Quantum Teleportation, Quantum Computation*. Springer, 2000, 314 pp.

[43] Chertov A.G. *Fizicheskiye Velichiny*. [Physical Quantities]. Russian. Moscow: Vyshshaya shkola, 1990, 335 pp.

[44] Dirac P.A.M. *The Principles of Quantum Mechanics*. Oxford: At the Clarendon Press, 1947.

[45] Lyubimov D.N. Dolgopolov K.N. *Fundamentalnyye osnovy evolutsii tribosistem*. [Fundamental Basics of Tribosystems Evolution]. Russian. Shakhty: FGBOU VPO "YURGUES", 2011, 95 pp.

[46] Lyubarsky G.S. "Itak ves khor ukazyvayet na taynyy zakon". [So the Entire Chorus Indicates One Secret Law]. Russian. In: *Znaniye-sila* 10 (1991), pp. 34–41.

[47] Born M. *Physics in My Generation.* Pergamon Press, 1956.

[48] Feynman R. Leighton R. Sands M. *The Feynman Lectures on Physics. Quantum Mechanics.* Vol. 3, Addison-Wesley, 1964.

[49] Medvedev B.V. *Nachala teoreticheskoy fiziki.* [Principles of Theoretical Physics]. Russian. Moscow: Nauka, 1977, 496 pp.

[50] Fermi E. *Notes on Quantum Mechanics.* Chicago: Univ. of Chicago Press, 1995, 191 pp.

[51] Aleksandrov P.S. et al., ed. *Encyklopedia Elementarnoy Matematiki.* [Encyclopedia of Elementary Mathematics]. Vol. 2: *Algebra.* Russian. Moscow: State Publishing House of Scientific and Technical Literature, 1954, 418 pp.

[52] Prokhorov A.M., ed. *Fizicheskaya encyklopedia.* [Physical Encyclopedia]. Russian. Vol. 1, Moscow: Sovetskaya enciklopedia, 1988, 704 pp.

[53] Barannikov A.V. Firsov A.V. *Osnovnyye kontseptsii sovremennoy fiziki.* [Basic Conceptions of Modern Physics]. Russian. Moscow: Vysshaya shkola, 2009, 390 pp.

[54] Feynman R. *QED—The Strange Theory of Light and Matter.* Princeton: Princeton Unvi. Press, 1983, 163 pp.

[55] Landau L.D. Lifschitz E.M. *Teoreticheskaya fizika.* [Theoretical Physics]. Vol. 2: *Teoriya pola.* [Field Theory]. Russian. Moscow: Nauka, 1974, 421 pp.

[56] Gardner M. *The New Ambidextrous Universe: Symmetry and Asymmetry from Mirror Reflections to Superstrings.* New York: Freeman and Co., 1990, 392 pp.

[57] Lyubimov D.N. Dolgopolov K.N. *Treniye i teoriya otnositelnosti: Vremennyye anomalii v tribosistemakh.* [Friction and Theory of Relativity: Time Anomalies in Tribosystems]. Russian. Shakhty: FGBOU VPO "YURGUES", 2011, 126 pp.

[58] Landau L.D. Lifschitz E.M. *Teoreticheskaya fizika.* [Theoretical Physics]. Vol. 1: *Mekhanika.* [Mechanics]. Russian. Moscow: Nauka, 1973, 208 pp.

[59] Lyubimov D.N. Pinchuk L.S. Dolgopolov K.N. *Tribofizika.* [Tribophysics]. Russian. Rostov-on-Don: South Federal Univ. Press, 2011, 296 pp.

[60] Lyubimov D.N. Dolgopolov K.N. Vershinin N.K. "Triboaktiviruyemyy doplerovskiy sdvig". [Triboactivated Doppler shift]. Russian. In: *Treniye i smazka v mashinakh i mekhanizmakh. [Friction and Lubrication in Machines and Mechanisms]* 8 (2012), pp. 29–34.

[61] Lyubimov D.N. Pinchuk L.S. et al. "Eksperimentalnoye podtverzhdeniye relativistskoy prirody treniya". [Experimental Verification of Relativistic Nature of Friction]. Russian. In: *Vestnik Grodnenskogo gosudarstvennogo universiteta im. Yanki Kupaly. [Herald of Yanka Kupala Grodno State University]* 4 (2012(141)), pp. 54–66.

[62] Hawking S. *My Brief History.* New York: Bantam, 2013, 144 pp.

[63] Chetverukhin N.F. *Izobrazheniye figur v kurse geometrii.* [Representation of Figures in Geometry Course]. Russian. Moscow: Uchpedgiz, 1959, 213 pp.

[64] Cooper L. *Physics: Structure and Meaning.* Lebanon: Brown, 1992, 566 pp.

[65] M. Gardner. *Time Travel and Other Mathematical Bewilderments.* New York: Freeman and Co., 1988, 295 pp.

[66] Chichinadze A.V., ed. *Osnovy tribologii.* [Foundations of Tribology]. Russian. Moscow: Nauka i Tekhnika, 1995, 778 pp.

[67] Khaykin S.E. *Fizicheskiye osnovy mekhaniki.* [Physical Foundations of Mechanics]. Russian. Moscow: Nauka, 1971, 751 pp.

[68] Yelutin P.V. Krivchenkov V.D. *Kvantovaya mekhanika.* [Quantum Mechanics]. Russian. Moscow: Nauka, 1976, 325 pp.

[69] Sivukhin D.V. *Termodinamika i molekularnaya fizika.* [Thermodynamics and Molecular Physics]. Vol. 2: *Obshchiy kurs fiziki.* [General Course of Physics]. Russian. Moscow: Nauka, 1975, 552 pp.

[70] Karyakin N.I. Bystrov K.N. Kireyev P.S. *Kratkiy spravochnik po fizike.* [Brief Guide in Physics]. Russian. Moscow: Vyshshaya shkola, 1964, 567 pp.

[71] Lyubimov D.N. Dolgopolov K.N. Vershinin N.K. "Mikromekhanizmy obrazovaniya obolochechnoy struktury triboplazmy". [Micromechanisms of Formation of Triboplasma Shell Structure]. Russian. In: *Treniye i smazka v mashinakh i mekhanizmakh. [Friction and Lubrication in Machines and Mechanisms]* 7 (2012), pp. 10–14.

[72] Landau L.D. Lifschitz E.M. "Elektrodinamika sploshnykh sred". [Continuum Electrodynamics]. Russian. In: *Teoreticheskaya fizika.* [Theoretical Physics]. Vol. 8. Moscow: Fizmatlit, 2005.

[73] Prokhorov A.M., ed. *Fizicheskaya Enciklopediya.* [Physical Encyclopedia]. Russian. Vol. 2. Moscow: Sovetskaya Enciklopedia, 1990, 703 pp.

[74] Lyubimov D.N. Dolgopolov K.N. Pinchuk L.S. "Vliyaniye obmennogo vzaimodeystviya mezhdu chastitsami triboplazmy na yeye strukturu". [The Impact of Exchange Interaction between Triboplasma Particles on its Structure]. Russian. In: *Treniye i smazka v mashinakh i mekhanizmakh* 12 (2012), pp. 6–13.

[75] Arsenov O.O. *Fizika vremeni.* [Physics of Time]. Russian. Moscow: Eksmo, 2010, 224 pp.

[76] Voynov K.N., ed. *Tribologiya: mezhdunarmezhdu enciklopediya.* [Tribology: International Encyclopedia]. Russian. Vol. 1. Krasnodar: Anima, 2010, 176 pp.

[77] Lyubimov D.N. Pinchuk L.S. Dolgopolov K.N. "Nelokalnyye svyazi mezhdu kvantovymi podsistemami tribosistem". [Nonlocal Connections between Quantum Subsystems of Tribosystems]. Russian. In: *Vestnik Grodnenskogo gosudarstvennogo universiteta im. Yanki Kupaly. [Herald of Yanka Kupala Grodno State University]* 4 (2013(163)), pp. 48–58.

[78] Lyubimov D.N. Pinchuk L.S. Dolgopolov K.N. *Kvantovaya paradigma tribologii.* [Quantum Paradigm of Tribology]. Russian. Rostov-on-Don: South Federal Univ. Press, 2013, 206 pp.

Index